FEEDBACK AMPLIFIER PRINCIPLES

SOL ROSENSTARK

Electrical Engineering Department
New Jersey Institute of Technology

MACMILLAN PUBLISHING COMPANY
NEW YORK
COLLIER MACMILLAN PUBLISHERS
LONDON

Macmillan Publishing Company
866 Third Avenue, New York, NY 10022

Collier Macmillan Canada, Inc.

Printed in the United States of America

printing number
1 2 3 4 5 6 7 8 9 10

Library of Congress Cataloging in Publication Data

Rosenstark, Sol
Feedback amplifier principles.

Includes index.
1. Feedback amplifiers. I. Title.
TK7871.58.F4R67 1985 621.3815′35 85-18877
ISBN 0-02-947810-3

To Jeannie, Mike, and Dan

Contents

Preface

This book is intended for the electrical engineer interested in obtaining a background in the area of feedback amplifiers. A suitable text for self-study or formal course work, the entire book can be covered in one semester at the senior undergraduate level or at the first-year graduate level.

I have taught a one-term graduate course on this material at the New Jersey Institute of Technology for more than ten years. In the early days when it was a one-year course, it covered signal-flow diagrams, a great deal of complex-variable material (particularly the Hilbert transform and the gain-phase relations for minimum phase structures), and the Bode ideal loop-gain characteristic. After some time it was determined that signal-flow graphs are not essential for the understanding of feedback amplifiers and that Bode's mathematical ideal is very difficult to approximate; equally good results can be obtained by other much simpler techniques. As a result, the course was streamlined and revised to cover material that is indispensable for the understanding and design of properly functioning feedback amplifiers. The students, who were getting tired of copying my notes from the blackboard, felt that the material had become sufficiently well organized to be put into a book. Their encouragement was instrumental in getting the work started on the manuscript.

This book is meant to fill a void which has existed in this area for many years. The first authoritative book on the subject was *Network Analysis and Feedback Amplifier Design* by H. W. Bode (Van Nostrand, 1945). Anyone who has used that book knows that it covered the material quite thoroughly, but it was very difficult to read. A book which clarified Bode's material was *Feedback Circuit Analysis* by S. S. Hakim (London Iliffe Books Ltd., 1966). It was modern for its time, but a great deal of the material dealt with electron-tube applications. There are other books on the subject, but they are reference books, written for the experienced specialist and do not contain exercise problems. Most textbooks on electronics that have been published in recent years devote a chapter to this subject, leaving students confused on material that requires a much more thoroughgoing treatment. Some instructors feel that feedback amplifiers are a subset of control systems. This kind of treatment is inadequate, because in control systems it is assumed that most blocks comprising the system have one output variable related to one input variable and loading is not considered an important issue. In feedback amplifiers we have two variables at the output (voltage and current) and two variables at the input. Loading of the feedback circuit on the output and input of the amplifier can only be ignored in cases where a large impedance mismatch exists. The impedance mismatch assumption is not a requirement for the proper application of the theory of this book.

The material presented in this book is entirely consistent with the classical methods first elaborated by Bode. Bode's matrix formulations (arising from mesh and nodal equations) were very elegant, but at times tended to obscure cause-and-effect relationships between element values and final circuit performance. The methods used in this book are much more direct, so that the connection between component values and final circuit performance is obvious. The book contains many illustrative examples, so that readers should have no difficulty studying it at their own pace.

I have taken deliberate pains to reduce the amount of background material required for the reading of this book. Toward that end, the amount of circuit analysis required for calculations consists of the voltage-divider theorem, the current-divider theorem, Thevenin's theorem, and Norton's theorem. Mesh analysis is needed only once in connection with the *RC* phase-shift oscillator and even there its use can be avoided. The material in App. A gives very convenient methods of performing calculations on bipolar and field-effect transistor circuits, eliminating the need to resort to mesh or nodal analysis. In this way it is always easy to see which elements in the circuit control what quantities, and how to select them to meet a particular specification. This material is essential to the understanding of the methods of Chaps. 2–4 and should be read first if it seems unfamiliar. It was felt that this material is tutorial in nature and therefore was put into App. A so that the rest of the book could be arranged in a more logical order.

Aside from an understanding of the aforementioned topics, some background in frequency-response calculations in amplifier circuits is also desirable. A background in complex variables may be helpful but is not essential, and the reader who wishes to understand the derivation of the Nyquist stability criterion in Chap. 6 will find a short review of the subject in App. B.

The asymptotic gain formula presented in Chap. 2 makes it possible to analyze any feedback amplifier circuit, even the emitter follower, regardless of whether it is possible to discern a feedback loop. Chapter 2 also presents the Blackman formula for evaluating impedance in feedback amplifiers. This theorem has been largely ignored in most textbooks on electronics, but the fact remains that it gives a method for calculating impedances in an unequivocal manner, even in confusing situations.

Chapter 3 simply presents some results on well-known feedback amplifier configurations for the reader's convenience and also discusses the less popular, but at times useful, bridge feedback connection.

Chapter 4 takes up the question of the proper breaking of the feedback loop and performing correct loop-gain measurement, even in situations in which it is impossible to find a point in the amplifier where there is a substantial impedance mismatch. R. D. Middlebrook presented methods which are suitable for the measurement of loop gain in dc-coupled feedback amplifiers in which it is difficult to open the loop without disrupting the operation of the amplifier. A somewhat more convenient alternative method is presented in Chap. 4, so that loop-gain measurements can be performed without any doubt that the loop has been properly broken and properly terminated.

Chapter 5 concerns itself with the calculation of loop-gain frequency response. The use of the Miller effect is shown to be ineffective for feedback amplifier analysis because this method only gives accurate results up to the 3-dB frequency, whereas an accurate knowledge of the frequency response is needed beyond gain crossover. The reader who has access to good circuit analysis program such as PCAP, CORNAP, or SPICE might find this chapter of marginal interest, but some interesting fundamental points can still be learned in this chapter. The frequency analysis methods that are presented can be carried out by using a pencil and paper and a calculator. In this way the reader who has no ready access to a computer with the proper software can still obtain excellent results.

In Chap. 6, methods for feedback amplifier stability analysis are presented. The emphasis here is on the Nyquist criterion. The method of root-locus is omitted since it is more applicable to control systems.

Chapter 7 examines feedback amplifier stability from the point of view of the time and frequency response. Explicit methods of capacitive compensation are given for various situations encountered in practice.

Chapter 8 takes up the question of the feedback required to attain various design specifications. Chapter 9 concludes the book by presenting some common oscillator configurations.

I wish to express my gratitude to my colleague Prof. Joseph Frank for the many thorough discussions on the subject of electronics and feedback amplifiers. They were most helpful in elucidating many theoretical and practical aspects of the subject, and made it possible to create a clearer presentation for much of the material in this book.

Sol Rosenstark

1

Introduction to Feedback Theory

1.1 *Introduction*

Although feedback amplifiers have been in use for more than half a century [1], a unified method of feedback amplifier analysis has yet to be accepted by feedback amplifier designers. The first attempt to establish proper theoretical foundations for feedback amplifier analysis was made by Bode and Blackman in the 1940s [2, 3], but very little (sometimes none) of Bode's and Blackman's work can be found in most modern day texts on electronic circuits. This book attempts to fill this gap by presenting Bode's and Blackman's work in modern terms. It also includes later results, which give very clear procedures for measuring loop gain, and then the subject of feedback amplifier stability is addressed. In all cases, an attempt is made to present the simplest possible methods of analysis, which also meet strict criteria for accuracy.

This chapter serves as an introduction to the concepts of feedback theory. It should not be assumed that the equations appearing here are to be used for design purposes. Equations appropriate to feedback amplifier analysis will be found in the following chapters.

Appendix A is included as a review of bipolar-transistor and FET (field effect transistor) equivalent circuit theory. Although most electronic circuit designers are very likely to be familiar with such material, it is included so that the reader can assure himself that his point of view and the author's correspond on this point.

1.2 *Elementary Feedback Theory*

A feedback system is one in which the output signal is used in some way to modify the input signal to the system as shown in Fig. 1.1.

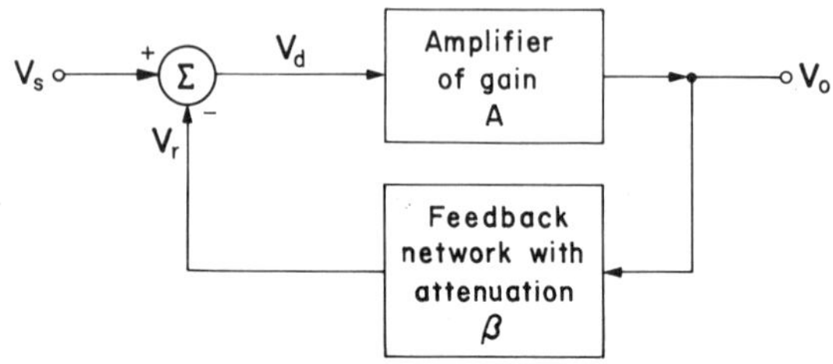

Figure 1.1 Feedback system block diagram.

In the model shown we have an amplifier with a forward gain A, which determines the signal relation between the output voltage V_o and the voltage at the input terminals V_d, according to

$$V_o = AV_d \tag{1.1}$$

In addition, we have a feedback network, which relates the returned signal V_r and the output signal V_o according to

$$V_r = \beta V_o \tag{1.2}$$

The summer takes the difference between the input signal V_s and the returned signal V_r to produce the difference signal

$$V_d = V_s - V_r \tag{1.3}$$

Substitution of (1.1) and (1.2) into (1.3) allows us to eliminate V_d and V_r from (1.3), leading directly to the gain with feedback,

$$A_f = \frac{V_o}{V_s} = \frac{A}{1 + \beta A} \tag{1.4}$$

The term βA which appears in the denominator of (1.4) is termed the *loop gain* or *return ratio* T of the feedback system. A method for determining the return ratio

$$T = \beta A \tag{1.5}$$

is shown in Fig. 1.2.

The signal source is set to 0. The signal path of the feedback loop is then broken at some convenient place and a signal of 1 unit is injected at terminal a. The returned signal at point b, V_{rb} is given by

$$V_{rb} = -\beta A \tag{1.6}$$

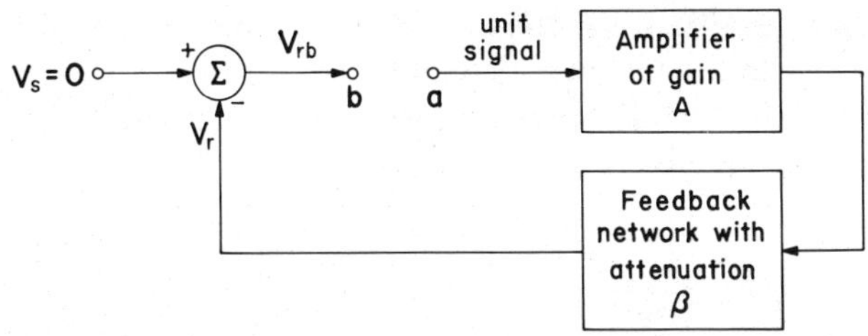

Figure 1.2 Feedback system prepared for finding T.

From this result and (1.5) we conclude that

$$T = -V_{rb} \tag{1.7}$$

Namely, the return ratio T is the *negative* of the returned signal at the output of the broken feedback loop, when a unit signal is injected at the input of the loop. The independent sources (V_s, in this example) must be set to zero before this procedure is carried out. We shall use the return ratio extensively later, so that the method of finding T will be of very great significance to us.

The signal appearing across terminals a–b is referred to as the *return difference* F, and we see from Fig. 1.2 that

$$F = 1 + \beta A \tag{1.8}$$

or

$$F = 1 + T \tag{1.9}$$

We can, in conclusion, write the gain with feedback in the form

$$A_f = \frac{A}{1 + T} \tag{1.10}$$

or

$$A_f = \frac{A}{F} \tag{1.11}$$

1.3 *Classification of Feedback Connections*

If the signal that is fed back is subtracted from the input signal, then the feedback is said to be *negative* or *degenerative*. If the feedback signal is added to the input signal, then the feedback is said to be *positive* or *regenerative*. For negative feedback, T is positive. Negative feedback is often used to obtain good performance from inferior equipment, the price being an increase in the quantity of equipment utilized. Sometimes a three-transistor feedback amplifier is built to obtain a good distortion specification, when the same gain could be obtained with a single-transistor amplifier, but with poor distortion performance. This point will be elaborated in the following sections.

1.4 *Sensitivity*

When we again examine (1.4), we observe that if the loop gain βA is much greater than unity, then

$$A_f \approx \frac{1}{\beta} \tag{1.12}$$

and the gain of the amplifier with feedback is entirely independent of variations in the gain A. This is the part of the feedback amplifier containing active elements, which vary with temperature, age, and component selection. Equation (1.12) states that if the return ratio is large then it is primarily the feedback network that controls the gain of the feedback amplifier. It is therefore important to make the feedback network independent of the kind of variations that active devices possess. Accordingly, the feedback network should be designed with stable passive devices.

If βA is not much greater than unity, we can still determine the sensitivity of the amplifier with respect to variations in A, by differentiating (1.4) with respect to A,

$$\frac{dA_f}{dA} = \frac{1}{(1 + \beta A)^2} \tag{1.13}$$

It is useful to define the sensitivity of the closed loop amplifier gain A_f as the fractional change in A_f for a given fractional change in A,

$$S_A = \frac{dA_f/A_f}{dA/A} \tag{1.14}$$

and we can readily obtain from (1.13), with the use of (1.5) and (1.10), the result,

$$S_A = \frac{1}{1 + T} \tag{1.15}$$

It can be readily appreciated, that for $T = 9$, a 10% change in the gain A will cause approximately a 1% change in the gain of the feedback amplifier. We must be mindful, however, that the gain variation of the amplifier is reduced by a factor of $1 + T$ at the expense of the amplifier gain, which has been reduced by the same factor.

Example 1.1: We are given an amplifier with a gain of 10 and a gain variation of 3.3%. The specifications call for an amplifier with a gain of 10 and a gain variation of 0.1%. We shall try to meet the specifications by putting three of the above amplifiers in cascade and using negative feedback. From the specifications we determine that

$$A = 10^3 = 1000$$

$$(\Delta A/A) = \{[10(1 + 0.033)]^3 - 10^3\}/10^3 = 0.1$$

$$A_f = 10$$

and

$$(\Delta A_f/A_f) = 0.001$$

Substituting this data into (1.14) and (1.15), we readily obtain the result $T = 99$, which is the required return ratio needed to fulfill the gain variation specification.

If we cascade three amplifier stages and then apply feedback, the resultant gain is

$$A_f = \frac{10^3}{1 + T} = \frac{10^3}{100} = 10$$

We see that we need three cascaded stages with feedback to meet the specification that one good stage could have met. Transistor prices being very low, one does not have to give a second thought to the idea of using many inferior stages where one good one could have done the job. ■

1.5 *Distortion*

Due to the nonlinear properties of the active devices used in amplifiers, distortion takes place which introduces extraneous signals into the amplifier output. As an example we will take a two-stage amplifier as shown in Fig. 1.3.

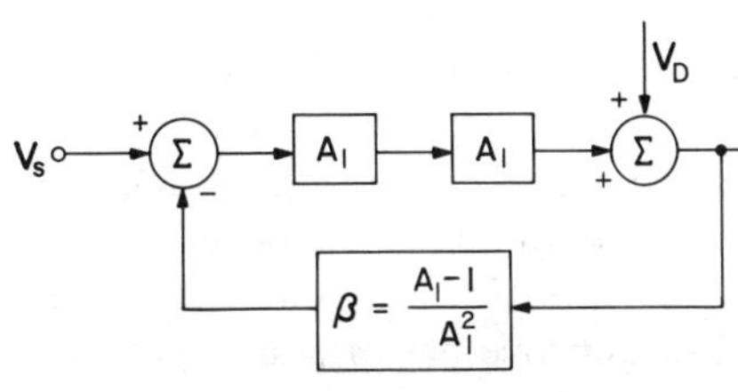

Figure 1.3 Two-stage amplifier for studying distortion.

The model shown represents a two-stage amplifier with the signal V_D representing the distortion at the amplifier output. The distortion signal appears at the output because distortion is generated by the stage that is driven the hardest.

We observe that the return ratio for the model shown is

$$T = \beta A_1^2 = A_1 - 1$$

Hence

$$1 + T = A_1$$

Applying (1.10) we see that when V_s is acting alone

$$\frac{V_o}{V_s} = \frac{A_1^2}{1 + T} = A_1$$

and when V_D is acting alone

$$\frac{V_o}{V_D} = \frac{1}{1 + T} = \frac{1}{A_1}$$

Using superposition we find that the total signal at the amplifier output is

$$V_o = A_1 V_s + \frac{1}{A_1} V_D \tag{1.16}$$

For comparison we choose an amplifier without feedback which has the same signal gain as that shown in Fig. 1.4.

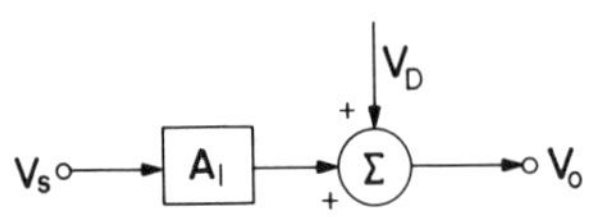

Figure 1.4 Nonfeedback amplifier used for comparison.

The output signal for this case is

$$V_o = A_1 V_s + V_D \tag{1.17}$$

and we see that the desired signal output is the same as that in (1.16), but the distortion in the feedback amplifier has been attenuated by a factor of A_1, which corresponds to the return difference of the feedback amplifier. Again we see that we can improve performance with inferior equipment if we use it in larger quantities. Sometimes more can be better.

1.6 *Noise*

We shall use the same models that were used to study harmonic distortion, except that we shall introduce noise at the input and output of each amplifier stage, as well as have noise accompany the signal at the amplifier input as shown in Fig. 1.5.

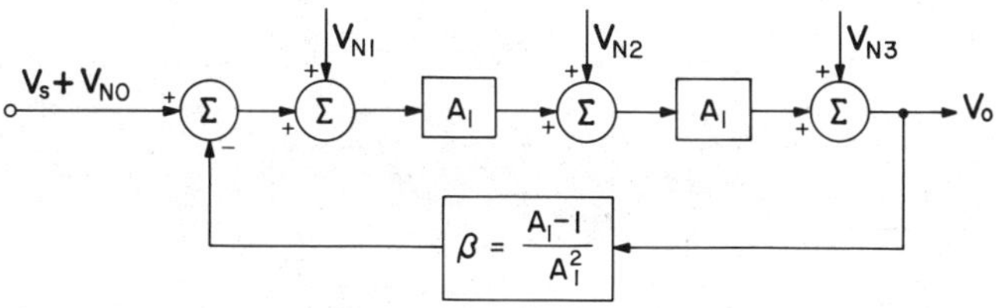

Figure 1.5 Two-stage amplifier used for studying the effects of additive noise.

Following the analysis of the previous example, we see that at the amplifier output we have

$$V_o = A_1(V_s + V_{NO} + V_{N1}) + V_{N2} + \frac{1}{A_1}V_{N3} \tag{1.18}$$

For comparison the nonfeedback amplifier is shown in Fig. 1.6.

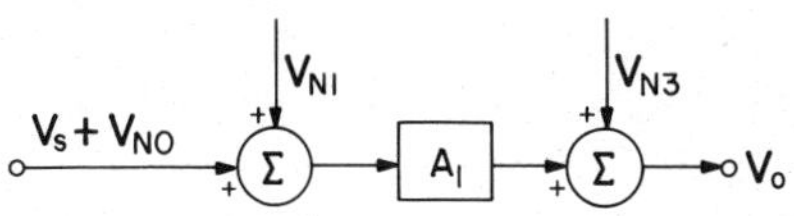

Figure 1.6 Nonfeedback amplifier used for comparison.

For this case, the total output voltage is

$$V_o = A_1(V_s + V_{NO} + V_{N1}) + V_{N3} \tag{1.19}$$

and we see that feedback does not give any improvement for noise introduced at the amplifier input, regardless of whether this noise source exists before or after the summer. But noise introduced close to the amplifier output is attenuated in the feedback amplifier, as in the case of a distortion signal. For cases where the greatest noise is introduced at the amplifier input, feedback gives no improvement at all.

1.7 *Bandwidth-Gain Trading*

The arguments which follow apply to amplifiers that possess a single 6-dB/octave high-frequency rolloff. In other words, we assume that the amplifier before feedback has a gain characteristic given by

$$A(f) = \frac{A_0}{1 + jf/f_h} \tag{1.20}$$

This is a low-frequency amplifier with a 20-dB/decade rolloff, with a break frequency f_h. We can trade gain for bandwidth by using the amplifier in a simple feedback configuration shown in Fig. 1.7.

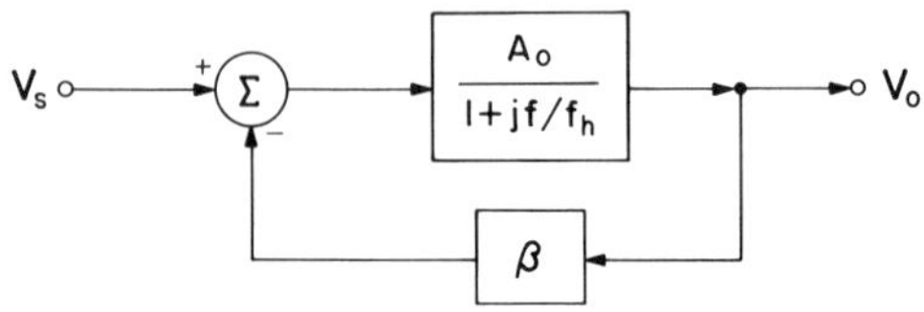

Figure 1.7 Feedback configuration for demonstrating gain-bandwidth trading.

The gain with feedback is found with the aid of (1.4)

$$A_f(f) = \frac{\dfrac{A_0}{1 + jf/f_h}}{1 + \beta \dfrac{A_0}{1 + jf/f_h}}$$

which results in

$$A_f(f) = \frac{\dfrac{A_0}{1 + \beta A_0}}{1 + j\dfrac{f}{f_h(1 + \beta A_0)}} \tag{1.21}$$

The above result can be written in the form

$$A_f(f) = \frac{A_0'}{1 + jf/f_h'} \tag{1.22}$$

where

$$A_0' = \frac{A_0}{1 + \beta A_0} \tag{1.23}$$

is the low-frequency gain reduced by feedback, and

$$f_h' = f_h(1 + \beta A_0) \tag{1.24}$$

is the 3-dB cutoff frequency in the presence of feedback. The gain-frequency tradeoff is very clear when the last two equations are examined.

Although the above analysis applies strictly to amplifiers possessing a single high-frequency pole, it could be generalized with some effort to amplifiers with more break frequencies. Furthermore, the same analysis can be carried out to show that gain-bandwidth trading applies to the low-frequency response of amplifiers just as well.

1.8 *Impedances*

Negative feedback can be used to change input and output impedances, a subject we shall explore thoroughly later in a unified approach for the various input–output configurations commonly found in feedback amplifiers.

1.9 *Conclusion*

The concepts presented in this chapter are no more than an introduction to feedback theory. Methods that are specifically applicable to feedback amplifiers will be presented in the following chapters. Almost all the equations presented in this chapter will be supplanted with other equations which are designed to give clear and unequivocal results even for such confusing feedback amplifier structures as the emitter follower.

At present there are two methods of arriving at feedback amplifier theory. One is the network formulation approach used by Bode [2] in laying down the fundamentals of feedback theory. His method was based on the use of mesh impedance or nodal admittance matrices. This approach is elegant, but is somewhat difficult to understand and lacks intuitive insight. The methods of feedback amplifier analysis, which will be presented in Chap. 2, are based on publications by Middlebrook [4] and Rosenstark [5]. The method used here relies on simple circuit analysis, which leads very naturally and directly to the feedback formulas, which will be used throughout the rest of this text. The theory presented will be found to be in entire agreement with that of Bode.

REFERENCES

1. Black, H. S.: Stabilized Feedback Amplifiers, *BSTJ*, vol. 13, January 1934, also U.S. Patent No. 2,102,671.
2. Bode, H. W.: "Network Analysis and Feedback Amplifier Design," Van Nostrand, New York, 1945.
3. Blackman, R. B.: Effect of Feedback on Impedance, *BSTJ*, vol. 22, pp. 269–277, October, 1943.
4. Middlebrook, R. D.: Design-Oriented Analysis of Feedback Amplifiers, Proc. Natl. Electronics Conf., pp. 234–238, October, 1964.
5. Rosenstark, S.: A Simplified Method of Feedback Amplifier Analysis, *IEEE Trans. on Ed.*, vol. 17, pp. 192–198, November 1974.

EXERCISES

1.1. It will be demonstrated that it is better to create a large cascade connection and put feedback on that, rather than putting feedback on individual stages.

Suppose we have an amplifier with a voltage transfer characteristic

$$A_0(f) = \frac{A_0}{1 + jf/f_0}$$

where

$$A_0 = 100$$

and

$$f_0 = 1 \text{ kHz}$$

(a) How much feedback (loop gain) is needed to change this amplifier's transfer characteristic to

$$A_1(f) = \frac{A_1}{1 + jf/f_1}$$

where $f_1 = 10$ kHz? What will be the value of A_1?

(b) Suppose four of the amplifiers whose gains are $A_1(f)$ are cascaded. What is the midband gain and 3-db frequency of the cascaded amplifier?

(c) The cascaded amplifier is heated so that A_0 changes by $+10\%$. By what percentage will the gain of the cascaded amplifier change?

(d) Now the alternate feedback amplifier is examined. Four of the stages whose gains are $A_0(f)$ are cascaded. How much loop gain is needed to get the midband gain of part (b)? What is the 3-db frequency of this connection? Compare the results to those of part (b).

(e) If A_0 changes by $+10\%$, how much will the overall amplifier gain change? Compare the results to those of part (c).

2

Fundamental Relations in Feedback Theory

2.1 *Fundamental Relations and Return Ratio*

Given any circuit, no matter how complicated, we can find the gain and input and output impedances by using any convenient circuit analysis technique such as mesh or nodal analysis. This kind of solution, however, hides the dependence of the computed parameters on the variation of the parameters of the active devices (e.g., transistors, FETs), which are contained in the circuit. We have found in the introduction that the loop gain T affects the stability of the gain of the feedback amplifier with respect to variations in the active-device gain A. It is therefore desirable to break up the solution of the feedback amplifier problem into lesser quantities, one of them being the return ratio T.

There is a difficulty in breaking up most feedback amplifiers into an active device with forward gain A, distinct from a feedback network with reverse attenuation β. The difficulty arises from the fact that all sections of the amplifier load one another, so that the input–output relation for each section cannot be represented in terms of a single parameter such as A or β. Accordingly, we shall define a quantity, the return ratio T, which is also the loop gain βA, without attempting to break up the feedback amplifier into distinct A and β networks. The entire feedback amplifier is left intact for these calculations.

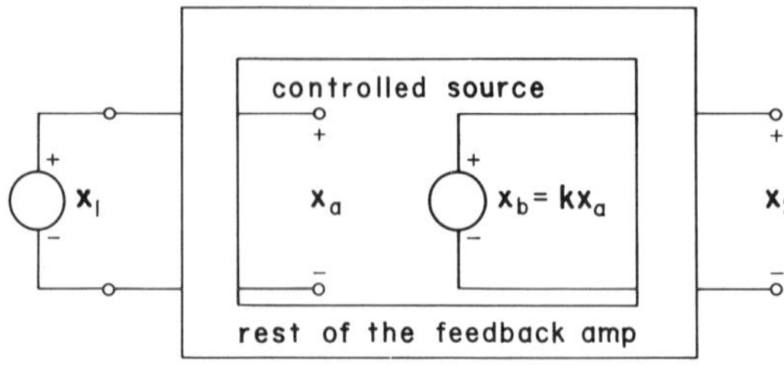

Figure 2.1 Feedback amplifier model.

The return ratio is calculated with reference to a specific controlled (dependent) source. Figure 2.1 shows the entire feedback amplifier contained in a rectangular box, which has input terminals excited by a source x_1 and output terminals showing an output x_2. There is a window in the box allowing us to see the reference controlled source x_b, whose dependence on the controlling quantity x_a is given by

$$x_b = kx_a \tag{2.1}$$

This controlled source could represent the current gain of a transistor, with x_a corresponding to the base current I_b, x_b corresponding to the collector current I_c, and k corresponding to the forward current gain h_f. It could, on the other hand, be identified with the parameters V_{gs}, V_{ds}, and μ of an FET. The fact that voltages are indicated in Fig. 2.1 should not lead to the conclusion that we are restricting ourselves to voltage analysis only. For want of a better method of representation, the quantities shown in Fig. 2.1 are voltages.

Definition. The return ratio T, with reference to controlled source x_b, is defined* as the negative of the variable x_a, which is produced when the dependent source x_b is replaced by an independent source of the same nature and polarity but of strength k, all independent sources are set to zero, and all other conditions in the system are left unchanged from their normal operating conditions.

To get an understanding of the definition, we shall calculate T in some examples.

Example 2.1: We shall find the return ratio T with reference to Q_2 for the amplifier shown in Fig. 2.2.

We first set the source V_i to 0, and then replace the dependent current source $h_{f2}I_{b2}$ with an independent current source of strength h_{f2}, and

*This method of defining return ratio was suggested by R. B. Blackman to D. E. Thomas, as is mentioned on page 1566 of [1].

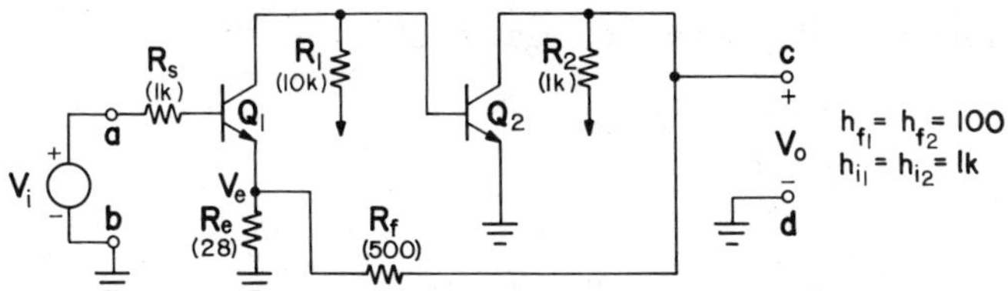

Figure 2.2 A two-transistor feedback amplifier.

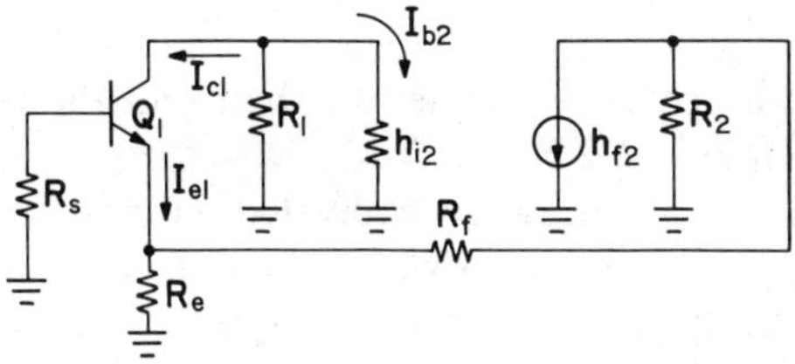

Figure 2.3 Equivalent circuit used for calculating T.

proceed to calculate I_{b2}. The negative of I_{b2} will be T. The resultant circuit is shown in Fig. 2.3.

Before proceeding, we replace Q_1 with another equivalent circuit according to the principles presented in App. A, so that the calculations can be made much simpler. As seen by an observer standing at the emitter of Q_1, all impedances above the emitter are reduced by a factor $(1 + h_f)$ as shown in Fig. 2.4. The part of the circuit needed for performing calculations involving I_{c1} is shown on the right.

First, we replace the circuit to the right of the dashed line with a Thevenin equivalent. Then it readily follows that

$$I_{e1} = \frac{h_{f2}R_2}{R_2 + R_f + R_e \| (h_{i1} + R_s)/(1 + h_{f1})} \times \frac{R_e}{R_e + (h_{i1} + R_s)/(1 + h_{f1})}$$

Since I_c is α times I_e, where $\alpha = h_f/(1 + h_f)$, it follows that

$$I_{b2} = -\alpha_1 I_{e1} \frac{R_1}{R_1 + h_{i2}}$$

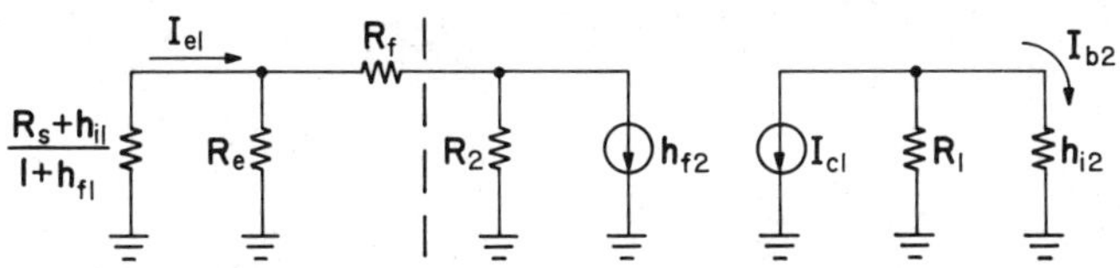

Figure 2.4 Simplified equivalent circuit.

The return ratio is merely the negative of I_{b2}, so that

$$T = \frac{\alpha_1 h_{f2} R_2}{R_2 + R_f + R_e \| (h_{i1} + R_s)/(1 + h_{f1})} \times \frac{R_e}{R_e + (h_{i1} + R_s)/(1 + h_{f1})} \frac{R_1}{R_1 + h_{i2}} \tag{2.2}$$

For the component values shown in Fig. 2.2, this evaluates to

$$T = 34.9$$

■

Example 2.2: For our second example, we shall find the return ratio for the amplifier shown in Fig. 2.5.

As in Example 2.1, we draw an equivalent circuit (according to the principles presented in App. A) in which the independent source V_i has been set to zero, and the dependent source $\mu_1 x_1$ has been replaced by an independent source of strength μ_1. The negative of the resultant x_1 is the desired value of T.

The equivalent circuit of Fig. 2.6 enables us to calculate V_1 in one step once we find the Thevenin equivalent of the circuit to the right of the dashed line. The resultant return ratio is given by

$$T = \mu_1 \frac{R_1}{r_{d1} + R_1} \frac{\mu_2}{1 + \mu_2} \frac{R_s}{R_s + (r_{d2} + R_2)/(1 + \mu_2)} \times \frac{R_g}{R_g + R_f + R_s \| (r_{d2} + R_2)/(1 + \mu_2)} \tag{2.3}$$

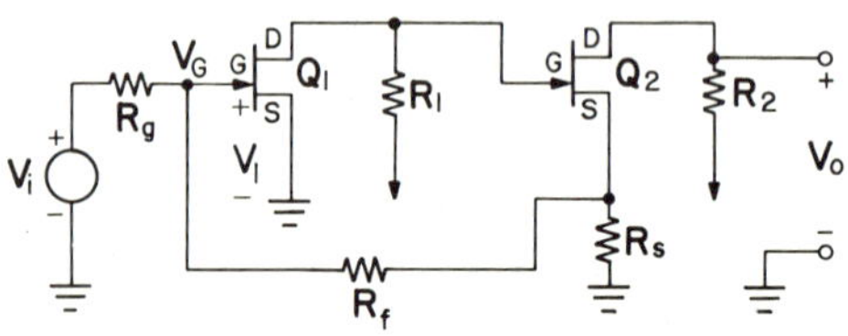

Figure 2.5 A two-FET feedback amplifier.

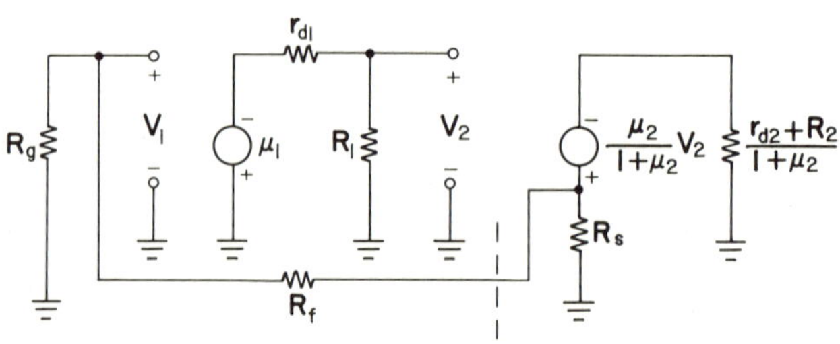

Figure 2.6 Equivalent circuit for calculating T.

We chose to calculate the return ratio in both examples with reference to transistors, which do not themselves have local feedback. Had we chosen to find the return ration with reference to the alternate transistor, we would have found that the resultant expressions are quite different, and in the case of a numerical example, we would have found that the return ratio is larger when calculated with respect to a transistor possessing local feedback in addition to overall amplifier feedback. ■

Now that we understand the definition for return ratio, and how to go about calculating it, we are ready to proceed to find complete expressions for the feedback amplifier gain.

2.2 *Characterization of Signal Transmission*

For Fig. 2.1 we already have (2.1) which relates x_b to x_a. We make the observation that setting x_b to zero can be accomplished by letting k take on the value of zero. We have to define some additional connective parameters before proceeding. We shall treat the two quantities x_2 and x_a as outputs, whereas x_1 and x_b will be treated as inputs. The equations connecting these four parameters are

$$x_a = G_{1a0}x_1 - G_{ba0}x_b \tag{2.4}$$

$$x_2 = G_{120}x_1 + G_{b20}x_b \tag{2.5}$$

The terms used in the above expressions have the following meaning:

$$G_{1a0} \equiv x_a \text{ produced by unit } x_1 \text{ when } k = 0 \tag{2.6}$$

$$G_{ba0} \equiv x_a \text{ produced by unit } x_b \text{ when } x_1 = 0 \tag{2.7}$$

$$G_{120} \equiv x_2 \text{ produced by unit } x_1 \text{ when } k = 0 \tag{2.8}$$

$$G_{b20} \equiv x_2 \text{ produced by unit } x_b \text{ when } x_1 = 0 \tag{2.9}$$

$$G_f \equiv x_2 \text{ produced by unit } x_1 \text{ when } k \neq 0 \tag{2.10}$$

The last expression is referred to as the external gain of the feedback amplifier or the feedback amplifier gain.

To find the return ratio T from the above expressions, we set x_1 to zero, let x_b take on the value k to find that the value returned to x_a is $-kG_{ba0}$. We see from this that

$$T = kG_{ba0} \tag{2.11}$$

We shall now make a first attempt at finding an expression for feedback amplifier gain.

2.3 *Asymptotic Gain Formula*

First we substitute (2.1) into (2.4) and (2.5) to obtain

$$0 = G_{1a0}x_1 - (1 + kG_{ba0})x_a \tag{2.12}$$

$$x_2 = G_{120}x_1 + kG_{b20}x_a \tag{2.13}$$

We solve the last two equations for

$$G_f = \frac{x_2}{x_1} \tag{2.14}$$

and perform enough algebraic manipulation to get the expression into the form

$$G_f = \frac{kG_{ba0}(G_{120} + G_{1a0}G_{b20}/G_{ba0}) + G_{120}}{1 + kG_{ba0}} \tag{2.15}$$

Using (2.11) in the above, along with the definitions

$$G_\infty = G_{120} + G_{1a0}G_{b20}/G_{ba0} \equiv \text{asymptotic gain} \tag{2.16}$$

$$G_0 = G_{120} \equiv \text{direct transmission term} \tag{2.17}$$

we get the *asymptotic gain relation*

$$G_f = G_\infty \frac{T}{1 + T} + \frac{G_0}{1 + T} \tag{2.18}$$

We observe that

$$G_f|_{T \to \infty} = G_f|_{k \to \infty} = G_\infty \tag{2.19}$$

and

$$G_f|_{T=0} = G_f|_{k=0} = G_0 \tag{2.20}$$

and it becomes clear why (2.18) is referred to as the asymptotic gain relation.

To compute G_∞ we need to allow $k \to \infty$ or $T \to \infty$. To see what condition this imposes on the circuit, we solve (2.12) for x_a

$$x_a = \frac{G_{1a0}}{1 + kG_{ba0}} x_1 = \frac{G_{1a0}}{1 + T} x_1 \tag{2.21}$$

and it becomes immediately apparent that

$$\lim_{k \to \infty} x_a = \lim_{T \to \infty} x_a = 0 \tag{2.22}$$

The condition $x_a = 0$ must be imposed on the circuit when calculating G_∞.

The method for calculating G_∞ and G_0 will now be demonstrated in two examples.

Example 2.3: We shall find G_∞ and G_0 with reference to Q_2 for the circuit in Fig. 2.2. To find G_∞ we impose on the circuit the condition of (2.22). Refer to Fig. 2.7 for the following steps.

Since h_{f2} is infinite then it follows that I_{b2} is 0. For this to be so, it is necessary for I_{c1} to be 0. Since h_{f1} is finite, it follows that I_{b1} and I_{e1} are also 0. As a consequence, there is no voltage drop across R_s and h_{i1}; therefore V_e must equal V_i. Since $I_{e1} = 0$, resistors R_f and R_e can be considered to be in series. Accordingly,

$$V_o \frac{R_e}{R_f + R_e} = V_e = V_i \tag{2.23}$$

Solving for V_o/V_i, we obtain

$$G_\infty = 1 + \frac{R_f}{R_e} \tag{2.24}$$

For the component values shown in Fig. 2.2, this evaluates to

$$G_\infty = 18.9$$

We calculated the above gain without ever having known the value of I_{c2}. This is finite but of no particular interest. It could be calculated now, if desired.

The direct transmission term is found from the circuit when $h_{f2} = 0$; hence I_{c2} is also 0. A circuit is drawn to give correct currents for an observer standing at the emitter of Q_1, when the methods of App. A are used. This is shown in Fig. 2.8.

It is found without great difficulty that

$$G_0 = \frac{V_o}{V_i} = \frac{R_e}{R_e + (R_s + h_{i1})/(1 + h_{f1})} \times \frac{R_2}{R_2 + R_f + R_e \| (R_s + h_{i1})/(1 + h_{f1})} \tag{2.25}$$

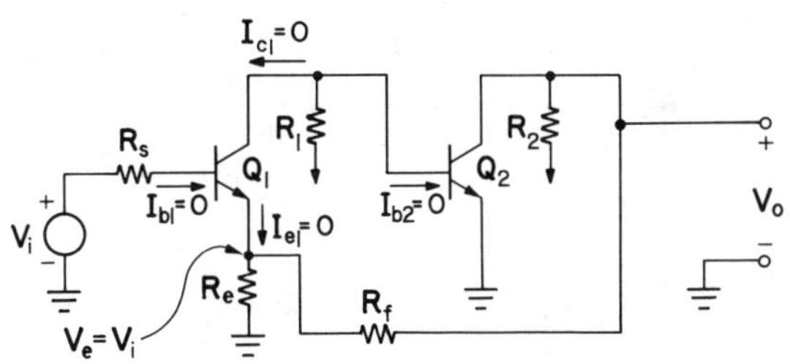

Figure 2.7 Circuit for finding G_∞.

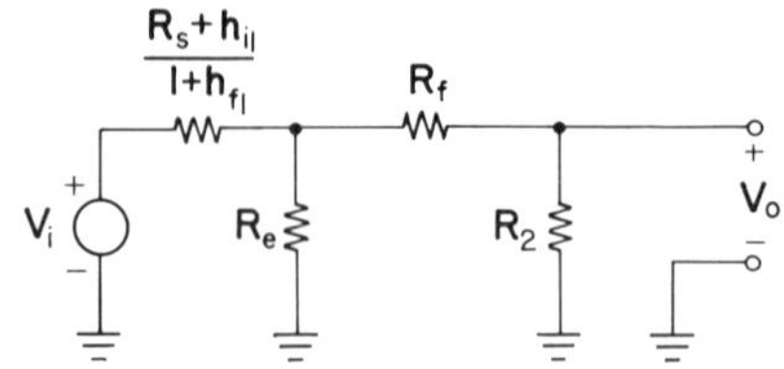

Figure 2.8 Equivalent circuit for finding G_0.

For the component values shown in Fig. 2.2, this evaluates to

$$G_0 = 0.388$$ ■

Example 2.4: We shall now find the asymptotic gain G_∞ with reference to Q_1 for the amplifier in Fig. 2.5. When μ_1 goes to infinity, V_1 goes to 0. The consequences of this are shown in Fig. 2.9.

Since $V_1 = 0$, then the current in R_g is V_i/R_g and the current in R_f is readily related to I_{d2} as shown in Fig. 2.9. The current from terminal a to b is zero. This would be the case regardless of the impedance connected from a to b since V_1 is zero. The currents in R_g and R_f must be equal and opposite, so it immediately follows that

$$I_{d2} = -\frac{V_i}{R_g}\frac{R_s + R_f}{R_s}$$

Since $V_o = -I_{d2}R_2$, we then conclude that

$$G_\infty = \frac{R_2}{R_g}\left(1 + \frac{R_f}{R_s}\right) \tag{2.26}$$

■

Examination of (2.18) shows that if

$$G_0 \ll G_\infty T \tag{2.27}$$

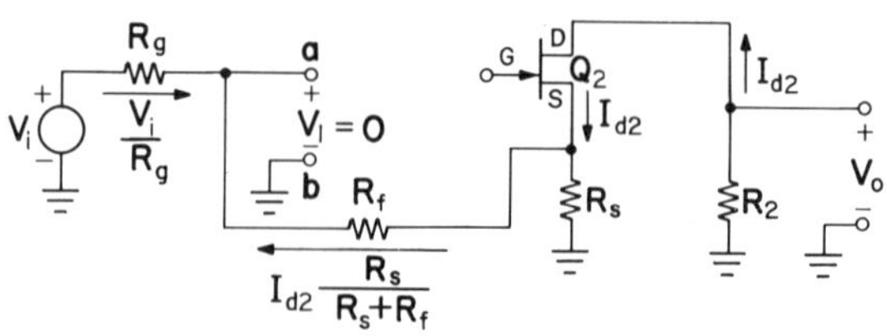

Figure 2.9 Circuit for finding G_∞ for the feedback amplifier of Fig. 2.5.

then the gain of the feedback amplifier is given very accurately by

$$G_f \approx G_\infty \frac{T}{1+T} \tag{2.28}$$

For the amplifier Example 2.1, (2.27) certainly holds, since $G_0 = 0.388$, $G_\infty = 18.9$, and $T = 34.9$, so that $G_\infty T = 1700G_0$. For certain amplifiers, however, G_∞ turns out to be 0, in which case G_0 is the only way to evaluate the gain of the amplifier when feedback is present. This is brought out in Prob. 2.7.

If, in addition to (2.27), $T \gg 1$ then G_∞ will represent the approximate external gain of the feedback amplifier. For the example given, saying that $G_f \approx G_\infty$ gives us a gain of 18.9, whereas using the more accurate (2.28), we get $G_f = 18.4$. The discrepancy between the two decreases as T becomes large compared to unity. It is important to remember that G_∞ gives a good first estimate of the closed loop gain of the feedback amplifier.

It is clear from the last example that the direct transmission term represents attenuation, so it will be less than 1. On the other hand, most feedback amplifiers will possess a return ratio T which will be much greater than 1. They will also have an external (hence asymptotic) gain which will be 1 or greater. It is clear that the inequality of (2.27) will hold in most cases, so that (2.28) becomes the expression for determining the external gain of most feedback amplifiers.

To put (2.28) into better perspective, we start from (1.4) and (1.5) to obtain

$$A_f = \frac{A}{1+\beta A} = \frac{1}{\beta}\frac{\beta A}{1+\beta A} = \frac{1}{\beta}\frac{T}{1+T}$$

We see that this last expression is incomplete when it is compared to (2.18), inasmuch as the G_0 term is completely disregarded. But by comparing this last expression to (2.28), we see that G_∞ and $1/\beta$ are identical. As was illustrated, one can easily calculate G_∞, whereas in most cases it is difficult to identify those parts of the feedback amplifier needed to calculate the attenuation β.

We see that using the asymptotic gain approach is no more difficult than using the A and β approach found in most control systems texts. Whereas in most control systems, the question of loading never comes up because the amplifier A is assumed to have a very low output impedance compared to the input impedance of the β network, the same cannot be said for feedback amplifiers. But this obstacle is easily surmounted when the asymptotic gain formula (2.18) is used. Only the two quantities T and G_∞ need be calculated in most cases, since G_0 will contribute very little to the overall feedback amplifier gain. The benefit of this method is that it is exact, and no approximations need be invoked before any of the quantities needed for the asymptotic gain formula are calculated.

Caution. In order to use the asymptotic gain formula correctly, the asymptotic gain G_∞, the return ratio T, and the direct transmission term G_0 must be calculated with reference to the same active device. It is usually easiest to perform all the calculations with reference to an active device which has a grounded emitter (for bipolar transistors) or a grounded source (for FETs).

2.4 *Blackman's Impedance Formula*

When it comes to calculating impedances in feedback amplifiers there is no easier method than the one using Blackman's formula [2], which will now be derived. More conventional techniques rely on identifying the type of feedback connection used (e.g., series–series, shunt–series, etc.) in order to determine whether one should divide or multiply by $1 + T$, the impedance seen at the terminals before feedback is applied. Blackman's formula dispenses with that requirement. One merely substitutes into the formula to obtain the result in a perfectly straightforward manner.

We shall refer to Fig. 2.10, which is almost identical to Fig. 2.1, except that the output x_2 now represents the voltage V at the amplifier terminals a–b, and the input x_1 now represents the current into the amplifier terminals a–b.

The relations characterizing Fig. 2.10 are

$$x_a = G_{1a0}I - G_{ba0}x_b \tag{2.29}$$

$$V = G_{120}I + G_{b20}x_b \tag{2.30}$$

and as always (2.1) relates x_b to x_a, so that setting x_b to 0 can be accomplished by setting k to 0. Before proceeding with the derivation a few definitions are in order.

$Z_{ab} \equiv$ Impedance seen at terminals a–b when the feedback amplifier is operating normally. (2.31)

$Z_{ab}^0 \equiv$ Impedance seen at terminals a–b when reference source $x_b = kx_a$ is 0. (Same condition as $k = 0$.) This is the null impedance at terminals a–b. (2.32)

$(T_{oc})_{ab} \equiv$ Return ratio with reference to source x_b when terminals a–b are open circuited. (2.33)

$(T_{sc})_{ab} \equiv$ Return ratio with reference to source x_b when terminals a–b are short circuited. (2.34)

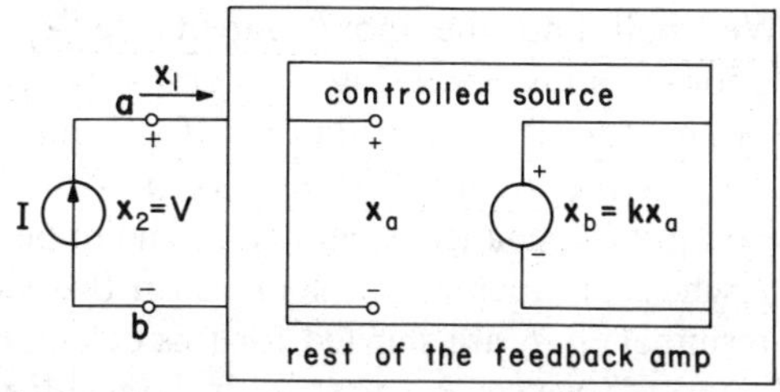

Figure 2.10 Feedback amplifier model for deriving Blackman's impedance formula.

We proceed by making the observation from (2.30) that

$$G_{120} = \left.\frac{V}{I}\right|_{x_b=0} = \left.\frac{V}{I}\right|_{k=0} = Z^0_{ab} \tag{2.35}$$

We shall now find expressions for $(T_{oc})_{ab}$ and $(T_{sc})_{ab}$. We follow the normal procedures for finding return ratio by replacing the controlled source x_b with an independent source of value k, and then proceeding to find x_a. The desired value of T is the negative of x_a. We restate (2.29) and (2.30) with x_b replaced by k,

$$x_a = G_{1a0}I - kG_{ba0} \tag{2.36}$$

$$V = G_{120}I + kG_{b20} \tag{2.37}$$

When terminals a–b are left open, I is zero and we find from (2.36) that

$$(T_{oc})_{ab} = kG_{ba0} \tag{2.38}$$

When terminals a–b are shorted, V is zero and we solve (2.36) and (2.37) for x_a whose negative is $(T_{sc})_{ab}$, with the result

$$(T_{sc})_{ab} = k(G_{ba0} + G_{1a0}G_{b20}/G_{120}) \tag{2.39}$$

Now that we are finished with the preliminaries, we find the expression for the impedance at terminals a–b when the feedback amplifier is normal. The expression obtained using (2.29), (2.30) along with (2.1) is

$$Z_{ab} = \frac{V}{I} = G_{120}\frac{1 + k(G_{ba0} + G_{1a0}G_{b20}/G_{120})}{1 + kG_{ba0}}$$

which can immediately be reinterpreted in terms of definitions (2.35), (2.38), and (2.39) into the form,

$$Z_{ab} = Z^0_{ab}\frac{1 + (T_{sc})_{ab}}{1 + (T_{oc})_{ab}} \tag{2.40}$$

This result is *Blackman's impedance formula*. Some examples will now be given for its use.

Example 2.5: We shall find the input impedance Z_i and the output impedance Z_o for the amplifier in Fig. 2.2. Transistor Q_2 will be used as the reference source for all the calculations. The input impedance is the impedance seen to the right of terminals a–b with the source V_i removed. The output impedance is defined as the impedance seen at the terminals c–d when all independent sources (in this case V_i) are set to zero, so that terminals a–b are shorted for this calculation.

To find Z_i^0 and Z_o^0 we set h_{f2} to 0, with the result that I_{c2} is also 0. Applying the methods of App. A on Fig. 2.2 we readily find that

$$Z_i^0 = R_s + h_{i1} + (1 + h_{f1})\left[R_e \| (R_f + R_2)\right] \tag{2.41}$$

and

$$Z_o^0 = R_2 \| \left[R_f + R_e \| (h_{i1} + R_s)/(1 + h_{f1})\right] \tag{2.42}$$

When the input terminals are shorted, the same conditions are obtained as those that existed when we calculated T in Example 2.1. It is therefore clear that

$$(T_{\text{sc}})_i = T \text{ of Eq. (2.2)} \tag{2.43}$$

When the input terminals are open, $I_{b1} = 0$, hence any voltage which is applied at the emitter has no effect on the collector current I_{c1}. As a consequence,

$$(T_{\text{oc}})_i = 0 \tag{2.44}$$

All that is left to be done is for (2.41), (2.43), and (2.44) to be substituted into Blackman's formula to calculate Z_i.

When the output terminals are shorted, any signal coming from Q_2 is shorted to ground, so that

$$(T_{\text{sc}})_o = 0 \tag{2.45}$$

When we attempt to calculate the return ratio with the output terminals left open, we see that we have the same conditions as existed when we calculated T in Example 2.1. Therefore

$$(T_{\text{oc}})_o = T \text{ of Eq. (2.2)} \tag{2.46}$$

As before all that is left to calculate the amplifier output impedance, is to substitute (2.42), (2.45), and (2.46) into Blackman's formula.

For the component values shown in Fig. 2.2, we get the results:

$Z_i^0 = 4.78\ \text{k}\Omega \qquad Z_i = 171\ \text{k}\Omega \qquad Z_o^0 = 338\ \Omega \qquad Z_o = 9.43\ \Omega$

We see that the effect of feedback was to raise the input impedance, and lower the output impedance of the feedback amplifier. We say this knowing full well that it is very difficult to define the input or output impedance of a feedback amplifier if feedback is absent. If R_f is removed in an effort to get rid of feedback then the input impedance and output impedance without feedback will not correspond to Z_i^0 and Z_o^0, respectively. This again points out the futility of attempting to model a feedback amplifier as a nonfeedback amplifier to which feedback has been added. Blackman's formula makes this kind of modelling entirely unnecessary. ■

Caution. In order to use Blackman's formula correctly, the null impedance Z^0, the short circuit return ratio T_{sc}, and the open circuit return ratio T_{oc} must be calculated with reference to the same active device. It is usually easiest to perform all the calculations with reference to an active device which has a grounded emitter (for bipolar transistors) or a grounded source (for FETs).

2.5 *Conclusion*

The techniques of dealing with feedback amplifier calculations presented in this chapter represent clear, unequivocal methods for finding all parameters of interest to the feedback amplifier designer. No approximations need be made before any of the calculations are carried out, so there is no need for prior "handwaving" before proceeding to solve feedback amplifier problems. The methods are very straightforward, and with some practice can be applied without any hesitation.

When performing gain calculations, a knowledge of the asymptotic gain G_∞ and the return ratio T are sufficient, since G_0 is very small in most cases. So only two quantities are needed to find the external gain G_f in most cases. The benefit of knowing G_∞ is that this represents a good approximation to the final external gain of the feedback amplifier. The return ratio T is needed in order to establish the amount of desensitization of the amplifier gain with respect to parameter variations, and in order to determine the amplifier's stability with respect to oscillation, as will be seen later.

Impedances are very easily calculated by using Blackman's formula. In most cases either T_{sc} or T_{oc} will be zero, and one of the two will usually (but not always) correspond to the return ratio T obtained when the gain calculations take place. We see that the only new quantity which has to be calculated is Z^0, and the amount of work involved is no greater than that needed when other methods are used.

REFERENCES

1. Thomas, D. E.: Some Design Considerations for High-Frequency Transistor Amplifiers, *BSTJ*, vol. 38, pp. 1551–1580, November, 1959.
2. Blackman, R. B.: Effect of Feedback on Impedance, *BSTJ*, vol. 22, pp. 269–277, October, 1943.

EXERCISES

2.1. For the one-transistor feedback amplifier shown in Fig. P2.1, find the return ratio T, the asymptotic gain G_∞, and the direct transmission term G_0.

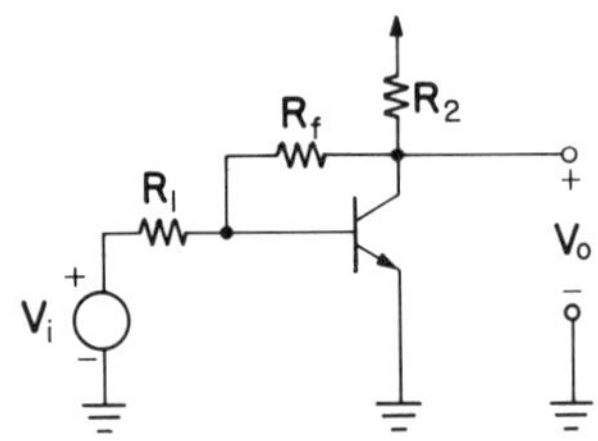

Figure P2.1

2.2. For the emitter follower shown in Fig. P2.2, find the return ratio T, the asymptotic gain G_∞, and the direct transmission term G_0. Substitute the resultant terms into the asymptotic gain formula, and see if the result is the same as would be obtained by conventional methods.

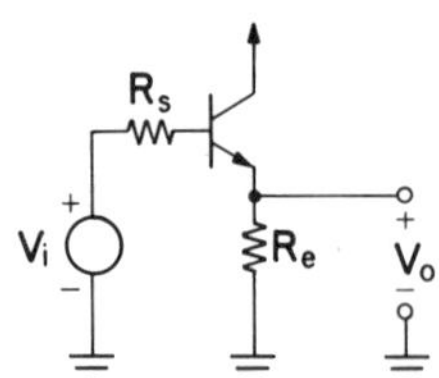

Figure P2.2

2.3. For the shunt–series two-transistor feedback amplifier shown in Fig. P2.3:

(a) Find the return ratio T, the asymptotic gain G_∞, and the direct transmission term G_0, with reference to Q_1.

(b) To convince yourself that it is easier to calculate all of the above quantities with respect to a grounded-emitter (or grounded-source) stage, calculate the return ratio with reference to Q_2.

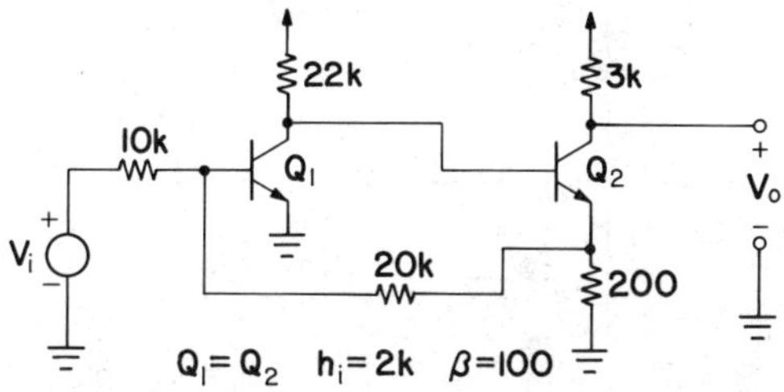

Figure P2.3

2.4. For the series–shunt two-FET feedback amplifier shown in Fig. P2.4, find the return ratio T, the asymptotic gain G_∞, and the direct transmission term G_0.

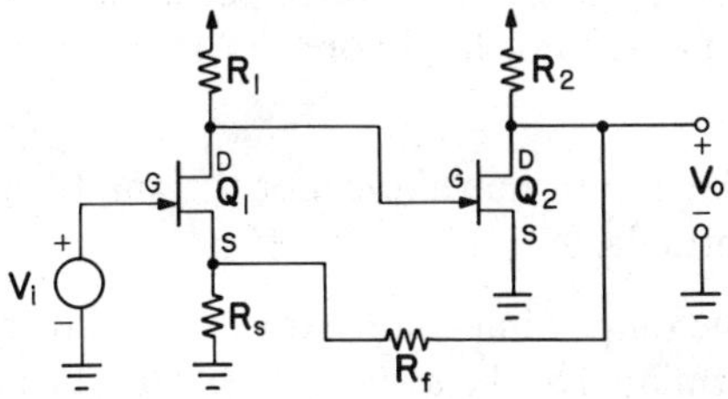

Figure P2.4

2.5. For the series–series, bipolar-transistor amplifier shown in Fig. P2.5, find the return ratio T, the asymptotic gain G_∞, and the direct transmission term G_0. Is inequality (2.27) satisfied?

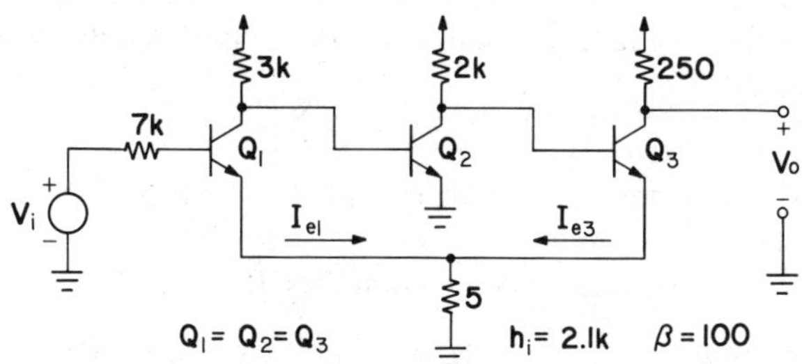

Figure P2.5

2.6. For the series–series FET feedback amplifier shown in Fig. P2.6, find the return ratio T, the asymptotic gain G_∞, and the direct transmission term G_0.

2.7. In Fig. P2.3 $V_i = 0$. There is a voltage source V_{r1} in series with the $22k$ collector resistor. This source represents power-supply ripple.

(a) Find the return ratio T, the asymptotic gain G_∞, and the direct transmission term G_0, which are needed in order to relate the output voltage V_o to the ripple voltage V_{r1}.

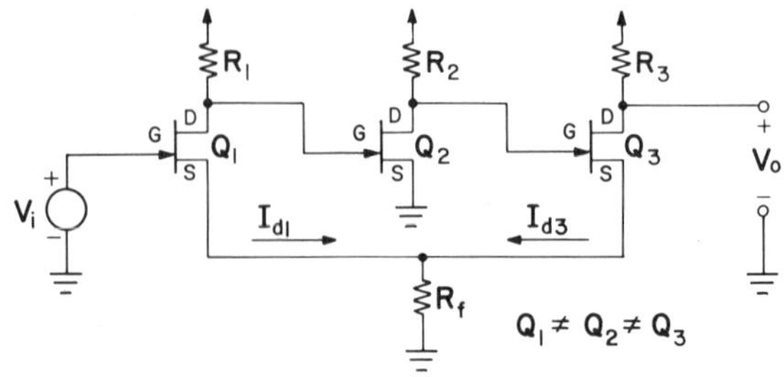

Figure P2.6

(b) Repeat this problem for a ripple source V_{r2} in series with the $3k$ resistor in Fig. P2.3. (Note: This problem demonstrates that there are cases where the asymptotic gain G_{∞} is zero, so the direct transmission term G_0 can definitely not be ignored in such situations.)

2.8. Find the input impedance and the output impedance for the amplifier of Fig. P2.1 using Blackman's formula.

2.9. Find the input impedance and the output impedance for the amplifier of Fig. P2.2 using Blackman's formula. Do the results agree with those that can be obtained if this is treated as a non-feedback problem and the solution is calculated by the impedance transformation methods of App. A?

2.10. Find the input impedance and the output impedance for the amplifier of Fig. P2.3 using Blackman's formula. Did feedback affect both impedances?

2.11. For the amplifier of Fig. P2.4:

(a) Find the input impedance and the output impedance using Blackman's formula.

(b) Now find the impedance seen across R_f.

2.12. Find the input impedance for the amplifier of Fig. P2.5 using Blackman's formula.

2.13. Find the output impedance for the amplifier of Fig. P2.6 using Blackman's formula.

3

Feedback Amplifier Connections

3.1 *Some Common Feedback Amplifier Connections*

There are many ways of using feedback in amplifiers to satisfy different operating requirements. The feedback connection has an effect on input and output impedance, and in addition, the connection determines whether an even number of stages or an odd number of amplifier stages have to be used. Some of the more common arrangements are the shunt input–shunt output, the shunt input–series output, series input–series output, and the series input–shunt output connections. In addition to the above, there is the less common bridge feedback connection. It is not necessary for us to know the classification for a particular amplifier, since the theory presented in the previous chapters does not require the amplifier to be put into a specific class before the feedback theory is applied. The names (such as shunt–shunt) are used to identify the amplifiers by a name which is commonly used in industry.

Figures 3.1–3.8 show eight common amplifier configurations, along with expressions for the return ratio T, the asymptotic gain G_∞, and the ratio of $G_\infty T/G_0$. The last quantity is given so that the significance of the contribution that G_0 makes to the asymptotic gain formula can be determined. In addition to the above quantities, the impedances and return ratios needed for substitution into Blackman's formula are also given. Using the expres-

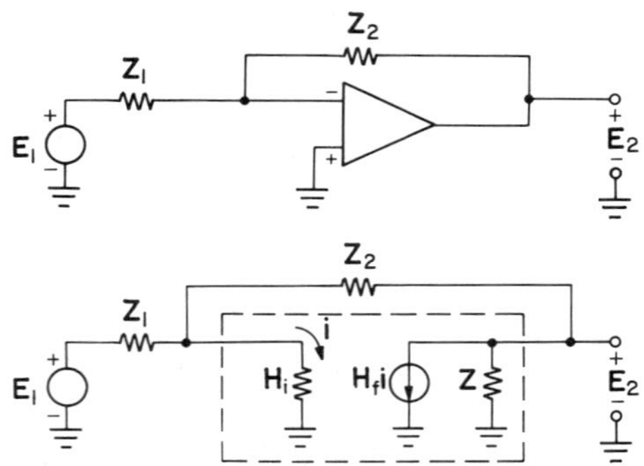

$$Z_1' = Z_1 \| H_i$$

$$T = \frac{H_f Z}{Z + Z_2 + Z_1'} \frac{Z_1}{Z_1 + H_i}$$

$$G_\infty = -\frac{Z_2}{Z_1} \qquad \frac{G_\infty T}{G_0} = -H_f \frac{Z_2}{H_i}$$

$$Z_{\text{in}}^0 = Z_1 + \left[H_i \| (Z_2 + Z) \right]$$

$$(T_{\text{sc}})_{\text{in}} = T \qquad (T_{\text{oc}})_{\text{in}} = T(Z_1 \to \infty)$$

$$Z_{\text{out}}^0 = Z \| [Z_2 + Z_1']$$

$$(T_{\text{sc}})_{\text{out}} = 0 \qquad (T_{\text{oc}})_{\text{out}} = T$$

Figure 3.1 Shunt–shunt feedback amplifier.

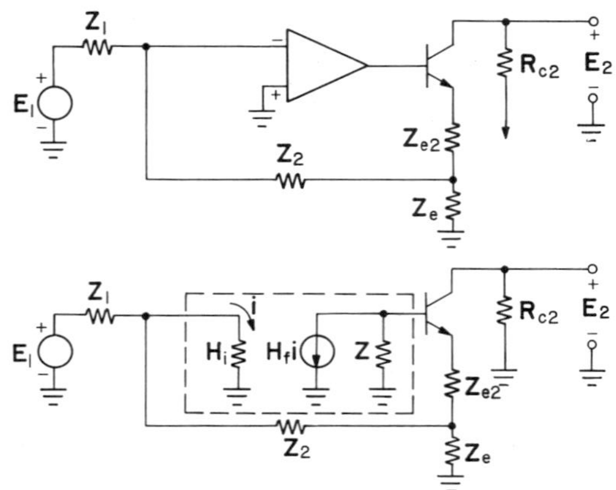

$$Z_2' = Z_2 + Z_1 \| H_i \qquad Z_{e2}' = Z_{e2} + \frac{r_{i2} + Z}{1 + \beta_2}$$

$$T = \frac{H_f Z}{Z_2' + Z_{e2}'(1 + Z_2'/Z_e)} \frac{Z_1}{H_i + Z_1}$$

$$G_\infty = \alpha_2 \frac{R_{c2}}{Z_1}\left[1 + \frac{Z_2}{Z_e}\right] \qquad \frac{G_\infty T}{G_0} = H_f \frac{Z}{H_i}\left[1 + \frac{Z_2}{Z_e}\right]$$

$$Z_{\text{in}}^0 = Z_1 + H_i \| (Z_2 + Z_e \| Z_{e2}')$$

$$(T_{\text{sc}})_{\text{in}} = T \qquad (T_{\text{oc}})_{\text{in}} = T(Z_1 \to \infty)$$

$$Z_{\text{out}}^0 = R_{c2}$$

$$(T_{\text{sc}})_{\text{out}} = T \qquad (T_{\text{oc}})_{\text{out}} = T$$

Figure 3.2 Shunt–series bipolar amplifier.

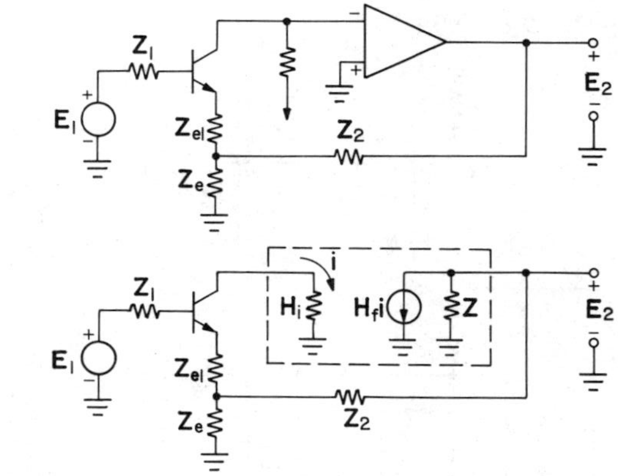

$$Z'_{e1} = Z_{e1} + \frac{r_{i1} + Z_1}{1 + \beta_1}$$

$$T = \frac{\alpha_1 H_f Z}{Z_2 + Z + Z'_{e1}[1 + (Z_2 + Z)/Z_e]}$$

$$G_\infty = 1 + \frac{Z_2}{Z_e} \qquad \frac{G_\infty T}{G_0} = \alpha_1 H_f \left[1 + \frac{Z_2}{Z_e}\right]$$

$$Z^0_{in} = (1 + \beta_1)[Z'_{e1} + Z_e \| (Z_2 + Z)]$$

$$(T_{sc})_{in} = T \qquad (T_{oc})_{in} = 0$$

$$Z^0_{out} = Z \| [Z_2 + Z_e \| Z'_{e1}]$$

$$(T_{sc})_{out} = 0 \qquad (T_{oc})_{out} = T$$

Figure 3.3 Series–shunt bipolar amplifier.

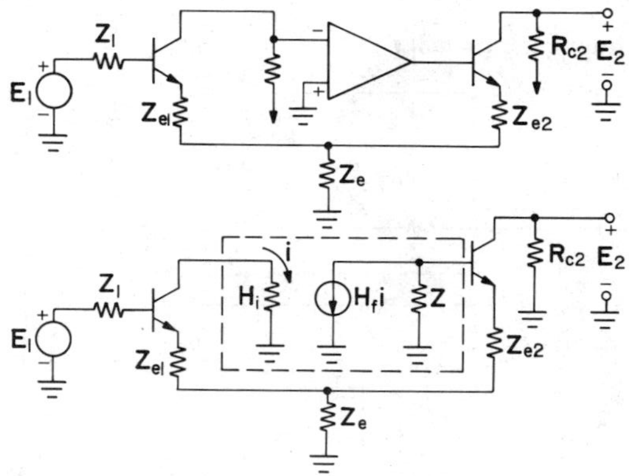

$$Z'_{e1} = Z_{e1} + \frac{r_{i1} + Z_1}{1 + \beta_1} \qquad Z'_{e2} = Z_{e2} + \frac{r_{i2} + Z}{1 + \beta_2}$$

$$T = \frac{\alpha_1 H_f Z}{Z'_{e1} + Z'_{e2}(1 + Z'_{e1}/Z_e)}$$

$$G_\infty = -\alpha_2 \frac{R_{c2}}{Z_e} \qquad \frac{G_\infty T}{G_0} = -\alpha_1 H_f \frac{Z}{Z_e}$$

$$Z^0_{in} = (1 + \beta_1)[Z'_{e1} + Z_e \| Z'_{e2}]$$

$$(T_{sc})_{in} = T \qquad (T_{oc})_{in} = 0$$

$$Z^0_{out} = R_{c2}$$

$$(T_{sc})_{out} = T \qquad (T_{oc})_{out} = T$$

Figure 3.4 Series–series bipolar amplifier.

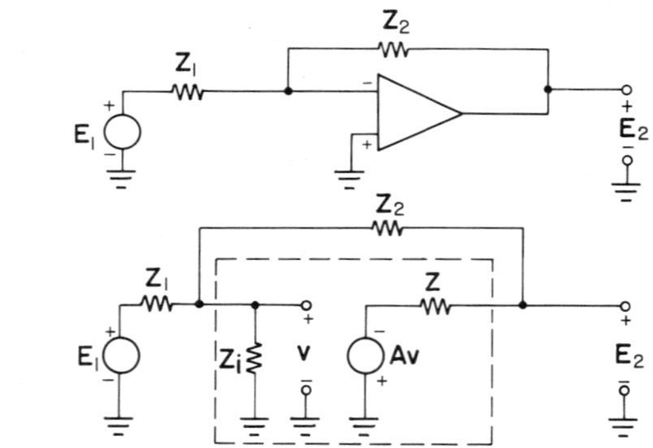

$$Z_1' = Z_1 \| Z_i$$

$$T = \frac{AZ_1'}{Z + Z_2 + Z_1'}$$

$$G_\infty = -\frac{Z_2}{Z_1} \qquad \frac{G_\infty T}{G_0} = -A\frac{Z_2}{Z}$$

$$Z_{\text{in}}^0 = Z_1 + [Z_i \| (Z_2 + Z)]$$

$$(T_{\text{sc}})_{\text{in}} = T \qquad (T_{\text{oc}})_{\text{in}} = T(Z_1 \to \infty)$$

$$Z_{\text{out}}^0 = Z \| [Z_2 + Z_1']$$

$$(T_{\text{sc}})_{\text{out}} = 0 \qquad (T_{\text{oc}})_{\text{out}} = T$$

Figure 3.5 Shunt–shunt feedback amplifier.

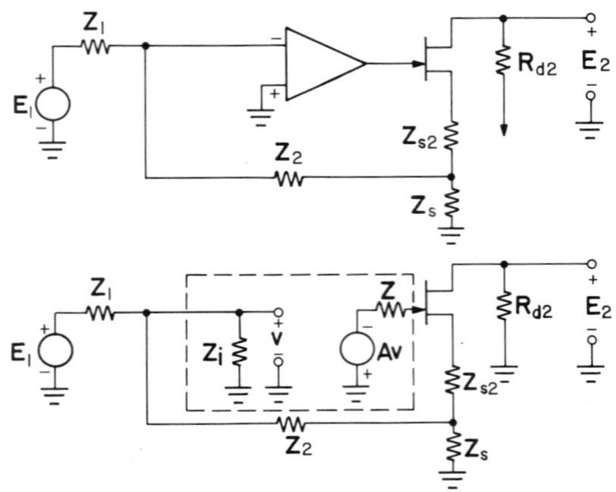

$$Z_1' = Z_1 \| Z_i \qquad Z_{s2}' = Z_{s2} + \frac{r_{d2} + R_{d2}}{1 + \mu_2}$$

$$T = \frac{\mu_2}{1 + \mu_2} \frac{AZ_1'}{Z_1' + Z_2 + Z_{s2}'[1 + (Z_1' + Z_2)/Z_s]}$$

$$G_\infty = \frac{R_{d2}}{Z_1}\left[1 + \frac{Z_2}{Z_s}\right] \qquad \frac{G_\infty T}{G_0} = A\frac{\mu_2}{1 + \mu_2}\left[1 + \frac{Z_2}{Z_s}\right]$$

$$Z_{\text{in}}^0 = Z_1 + Z_i \| (Z_2 + Z_s \| Z_{s2}')$$

$$(T_{\text{sc}})_{\text{in}} = T \qquad (T_{\text{oc}})_{\text{in}} = T(Z_1 \to \infty)$$

$$Z_{\text{out}}^0 = R_{d2} \| \{ r_{d2} + (1 + \mu_2)[Z_{s2} + Z_s \| (Z_2 + Z_1')] \}$$

$$(T_{\text{sc}})_{\text{out}} = T(R_{d2} = 0) \qquad (T_{\text{oc}})_{\text{out}} = T$$

Figure 3.6 Shunt–series FET amplifier.

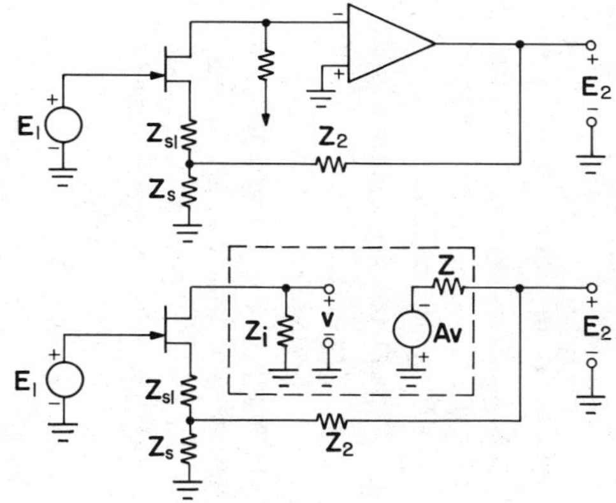

$$Z'_{s1} = Z_{s1} + \frac{r_{d1} + Z_i}{1 + \mu_1}$$

$$T = \frac{AZ_i}{Z + Z_2 + Z'_{s1}[1 + (Z + Z_2)/Z_s]}$$

$$G_\infty = \frac{\mu_1}{1 + \mu_1}\left[1 + \frac{Z_2}{Z_s}\right] \qquad \frac{G_\infty T}{G_0} = A\frac{Z_i}{Z}\left[1 + \frac{Z_2}{Z_s}\right]$$

$$Z_{\text{in}} = \infty$$

$$Z^0_{\text{out}} = Z \| (Z_2 + Z_s \| Z'_{s1})$$

$$(T_{\text{sc}})_{\text{out}} = 0 \qquad (T_{\text{oc}})_{\text{out}} = T$$

Figure 3.7 Series–shunt FET amplifier.

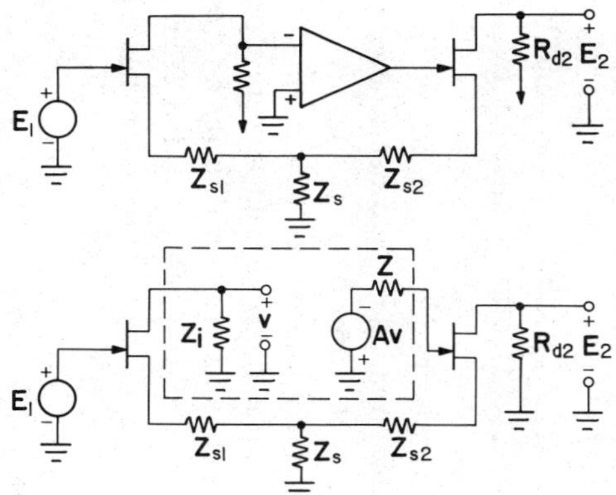

$$Z'_{s1} = Z_{s1} + \frac{r_{d1} + Z_i}{1 + \mu_1} \qquad Z'_{s2} = Z_{s2} + \frac{r_{d2} + R_{d2}}{1 + \mu_2}$$

$$T = \frac{\mu_2}{1 + \mu_2}\frac{AZ_i}{Z'_{s1} + Z'_{s2}[1 + Z'_{s1}/Z_s]}$$

$$G_\infty = -\frac{\mu_1}{1 + \mu_1}\frac{R_{d2}}{Z_s} \qquad \frac{G_\infty T}{G_0} = -A\frac{Z_i}{Z_s}\frac{\mu_2}{1 + \mu_2}$$

$$Z^0_{\text{out}} = R_{d2} \| [r_{d2} + (1 + \mu_2)(Z_{s2} + Z_s \| Z'_{s1})]$$

$$(T_{\text{sc}})_{\text{out}} = T(R_{d2} = 0) \qquad (T_{\text{oc}})_{\text{out}} = T$$

Figure 3.8 Series–series FET amplifier.

sions presented, it is easy to see that the input and output impedance is reduced when shunt–shunt feedback is used in an amplifier.

The low-frequency equivalent circuits used for the bipolar transistors and FETs are the same as those discussed in Appendix A. The techniques of Appendix A are very helpful in arriving at most of these expressions with a minimum of fuss. The only change in common notation was to substitute the designation r_i in place of h_{ie} and to use β in place of h_{fe}.

To apply the expressions shown in these diagrams, it is necessary to reduce the triangular inverting amplifier element to an equivalent form containing an input impedance, an output impedance, and a controlled source. The element shown as an inverting amplifier can be an integrated amplifier, or a discrete amplifier utilizing one, three, or five bipolar transistors, FETs, or a combination of those.

The amplifiers analyzed are presented in two groups. The first group is more suitable for performing calculations on amplifiers containing bipolar transistors, since they are more readily modelled in terms of controlled current sources. The second group presented is more suitable for analyzing FET amplifiers, since these are more readily modelled in terms of controlled voltage sources.

3.2 *Bridge Feedback*

We have seen that for the shunt and series feedback amplifier connections the return ratio is affected by the source and load impedances. If an amplifier is needed for a broad range of input or output impedances, then it is not desirable to have the return ratio dependent on source impedance or load impedance. Another characteristic of the typical feedback amplifiers examined previously is that the input or output impedance is directly or inversely related to the return ratio. If there are any variations in return ratio due to component selection or component aging, the input and output impedances will be directly affected. Another design disadvantage of the shunt- and series-connected feedback configurations is that since return ratio is affected by the amplifier terminations, its stability becomes dependent on the source and load impedances. These difficulties are overcome by the use of the bridge feedback arrangement shown in Fig. 3.9.

In Fig. 3.9 the impedance Z_1 incorporates the amplifier output impedance, and the impedance Z_b incorporates the amplifier input impedance. If both the input and output bridges are balanced, then any signal that is returned to the input by means of a voltage developed across terminals 4–5 is not affected by anything that is connected across terminals 2–2′. Therefore the return ratio is not affected by the load impedance Z_L. Hence

$$(T_{sc})_{22'} = (T_{oc})_{22'} \tag{3.1}$$

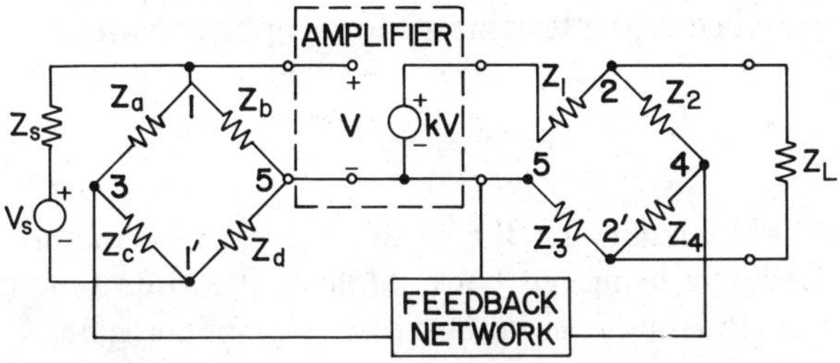

Figure 3.9 Bridge feedback amplifier connection.

and from Blackman's formula it then follows that

$$Z_{22'} = Z^0_{22'} \tag{3.2}$$

Hence feedback leaves the output impedance unaltered from the impedance seen at terminals 2–2′ when $k = 0$. A similar argument can be made for the input terminals 1–1′, so that the return ratio is not affected by the source impedance Z_S. The feedback amplifier input impedance is also unaltered by the presence of feedback.

If the bridges are perfectly balanced, then it is apparent from Fig. 3.9 that we can express the return ratio as

$$T_{sc} = T_{oc} = T = kK_0 \tag{3.3}$$

where K_0 is a quantity independent of k. If the bridges are not balanced, then the return ratio becomes dependent on how the input or output terminals are terminated, and we can write

$$T_{sc} = kK_{sc} \tag{3.4}$$

and

$$T_{oc} = kK_{oc} \tag{3.5}$$

The above can be used in Blackman's impedance relation, so that we can write

$$Z = Z^0 \frac{1 + kK_{sc}}{1 + kK_{oc}} \tag{3.6}$$

If the amplifier has a large loop gain, so that

$$T_{sc} = kK_{sc} \gg 1 \tag{3.7}$$

and

$$T_{oc} = kK_{oc} \gg 1 \tag{3.8}$$

then the impedance at the terminals can be approximated by

$$Z \approx Z^0 \frac{K_{sc}}{K_{oc}} \tag{3.9}$$

We see that unbalancing the bridges gives us a method of changing impedance levels by using feedback, while at the same time they are made relatively independent of variations of the amplifier gain k. This can be contrasted with the impedance for the series feedback case

$$Z = Z^0(1 + T_{sc}) = Z^0(1 + kK_{sc}) \approx Z^0 kK_{sc} \tag{3.10}$$

and with the admittance for the shunt feedback case

$$Y = Y^0(1 + T_{oc}) = Y^0(1 + kK_{oc}) \approx Y^0 kK_{oc} \tag{3.11}$$

in which the impedance (admittance) parameters vary directly with amplifier-gain variations.

Example 3.1: It is not immediately apparent that the amplifier shown in Fig. 3.10 has bridge feedback at the output. It does not need bridge feedback at the input because the source impedance has no effect on return ratio, and the input impedance of the amplifier is so high, that operation is not affected by R_s, the output impedance of the source V_i.

It will be assumed that all bypass capacitors are short circuits, the radio frequency choke RFC is an open circuit, and the 1-MΩ resistors can be left out of the equivalent circuit without affecting the analysis significantly. For finding output impedance, we set all independent sources to zero, which, in this case, means that the voltage source V_i is replaced by a short circuit.

The first equivalent circuit of the amplifier is shown in Fig. 3.11. The techniques of App. A were used to replace the first FET with an equivalent circuit from the point of view of an observer standing at the source terminal.

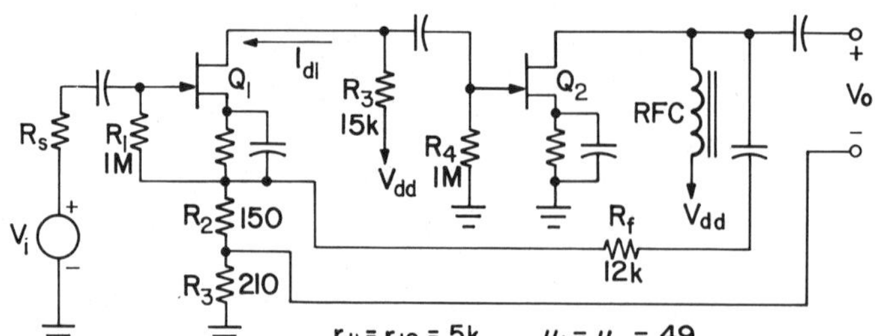

Figure 3.10 Amplifier with bridge feedback at the output.

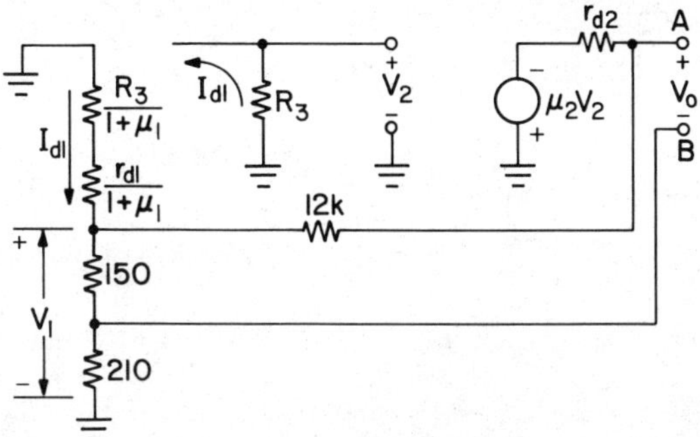

Figure 3.11 First equivalent circuit of the amplifier.

We now compute

$$\frac{r_{d1} + R_3}{1 + \mu_1} = \frac{5k + 15k}{1 + 49} = 400\ \Omega$$

$$I_{d1} = -V_1/400$$

$$V_2 = -R_3 I_{d1} = -(15k) I_{d1} = 37.5 V_1$$

$$\mu_2 V_2 = 49(37.5 V_1) = 1838 V_1$$

The new equivalent circuit is shown in Fig. 3.12. The fact that the amplifier has bridge feedback is now clear, but it is also apparent that the bridge is not balanced.

We are interested in finding the output impedance Z_{AB}. This is done by calculating the three quantities needed for substitution into Blackman's formula. The open-circuit return ratio is easily found from Fig. 3.12.

$$(T_{oc})_{AB} = \frac{(150 + 210)\|400}{(150 + 210)\|400 + 12k + 5k} 1838 = 20.25 \quad (3.12)$$

To calculate the short-circuit return ratio, we use the equivalent circuit of Fig. 3.13 which reflects the fact that terminal A is shorted to terminal B.

The short-circuit return ratio is readily found from Fig. 3.13.

$$(T_{sc})_{AB} = 1838 \frac{210}{5k + 210} \frac{400}{400 + 150\|12k + 210\|5k} = 39.5 \quad (3.13)$$

To find Z^0 we set the source $1838V_1$ to 0 as shown in Fig. 3.14*a*. To carry out the calculation we perform a $\Delta - Y$ transformation, with the result shown in Fig. 3.14*b*.

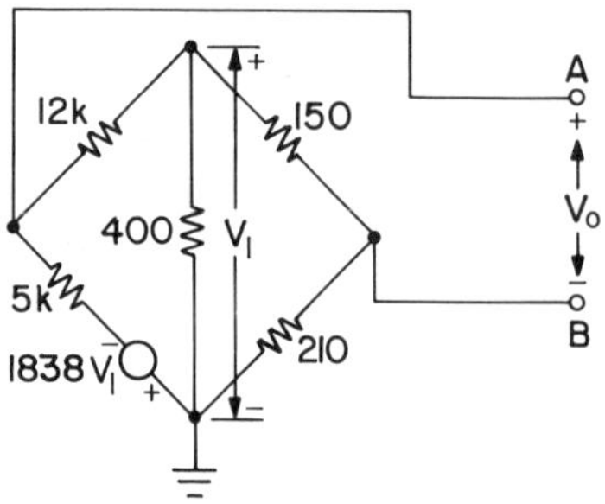

Figure 3.12 Equivalent circuit reduced to a bridge.

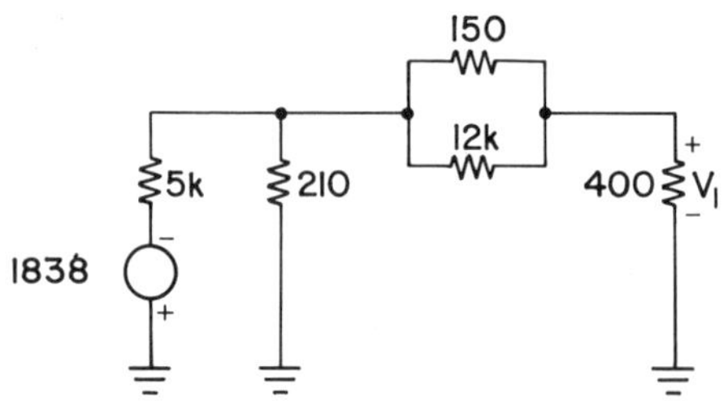

Figure 3.13 Equivalent circuit for finding T_{sc}.

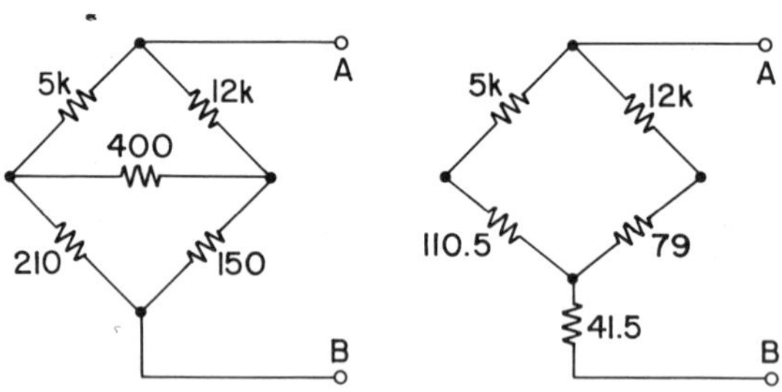

Figure 3.14 Equivalent circuit for finding Z^0_{AB}.

We can now easily calculate

$$Z^0_{AB} = 3.63k \tag{3.14}$$

The results in (3.12)–(3.14) can be used in Blackman's formula to calculate

$$Z_{AB} = Z^0_{AB}\frac{1+(T_{sc})_{AB}}{1+(T_{oc})_{AB}} = Z^0_{AB}\frac{1+39.5}{1+20.25} = 1.906Z^0_{AB} \tag{3.15}$$

If for some reason the gain of the second FET were to double, then Z^0_{AB} would be unchanged but $(T_{sc})_{AB}$ and $(T_{oc})_{AB}$ would both double, and the new value for Z_{AB} would become

$$Z_{AB} = Z^0_{AB}\frac{1+79}{1+40.5} = 1.928Z^0_{AB} \tag{3.16}$$

This represents a change in output impedance of 1.15%. In the case of a more conventional feedback connection there would have been a factor of 2 change in output impedance. ■

EXERCISES

3.1. Derive all the results which appear in Fig. 3.1.

3.2. Derive all the results which appear in Fig. 3.2.

3.3. Derive all the results which appear in Fig. 3.3.

3.4. Derive all the results which appear in Fig. 3.4.

3.5. Derive all the results which appear in Fig. 3.5.

3.6. Derive all the results which appear in Fig. 3.6.

3.7. Derive all the results which appear in Fig. 3.7.

3.8. Derive all the results which appear in Fig. 3.8.

3.9. Find the asymptotic gain G_∞ for the bridge feedback amplifier of Fig. 3.10.

3.10. Find the direct transmission term G_0 for the bridge feedback amplifier of Fig. 3.10.

4

Loop-Gain Measurement

PREVIOUS CHAPTERS HAVE DEALT with the calculation of loop gain (return ratio) from the schematic diagram of the feedback amplifier. Now we wish to examine methods of loop-gain measurement and the problems that arise in their application. The methods involving the breaking of the loop have always been in common use, and we will merely attempt to clarify their basis and see what inaccuracies are introduced if some simplifications are made to the circuits before measurements are made. Other methods will then be presented, which are more tedious to carry out, but are useful for amplifiers with special requirements.

4.1 *Breaking Loop—Voltage Ratio Measurement*

Any feedback amplifier can be drawn in the kind of diagram shown in Fig. 4.1. All passive and active elements have been collected into the block within the dashed lines. The only element which is external to this block is the feedback connection, shown here as a solid wire. The feedback connection will be broken between the terminals marked c–c'.

The return ratio calculated from Fig. 4.1 is

$$T = G_m \frac{Z_1 Z_2}{Z_1 + Z_2} \tag{4.1}$$

The loop is broken at c–c', and the cut end at c' is terminated with the

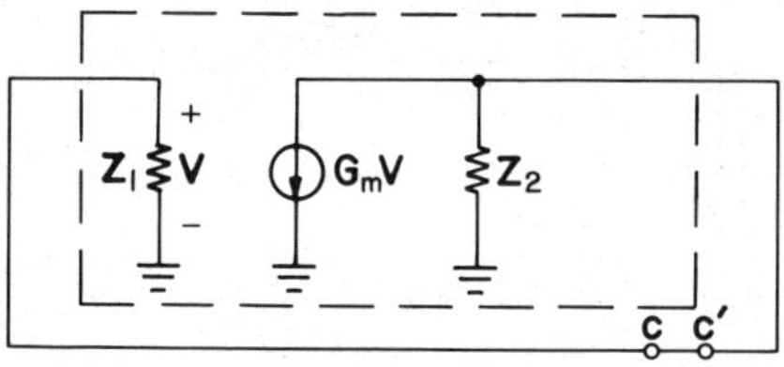

Figure 4.1 Feedback amplifier model.

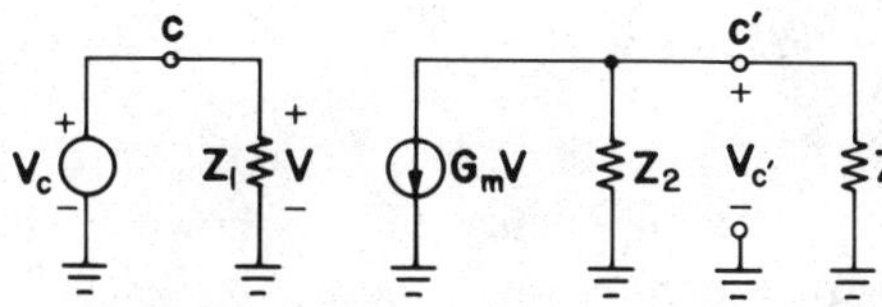

Figure 4.2 Amplifier with feedback loop cut and terminated.

impedance Z_1 as shown in Fig. 4.2. A voltage source V_c is connected at terminal c.

The ratio of $V_{c'}/V_c$ is calculated and it is found that

$$\frac{V_{c'}}{V_c} = -G_m \frac{Z_1 Z_2}{Z_1 + Z_2} \tag{4.2}$$

It is therefore apparent that we can obtain the return ratio by cutting and terminating the feedback loop and calculating the return ratio using

$$T = -\frac{V_{c'}}{V_c} \tag{4.3}$$

Clearly, we have to terminate at the cut with the impedance that the signal would encounter if it continued around the loop. If the cut is chosen so that the impedances on both sides of the cut are such that Z_1 is large compared to Z_2, then Z_1 can be replaced by an open circuit, but an inaccuracy will be introduced into the measurement. To find the error in the measurement, we observe that when T is calculated using (4.2) and (4.3) and Z_1 is replaced by an open circuit, the approximate return ratio using the voltage ratio is

$$T_{\text{oc}} = G_m Z_2 \tag{4.4}$$

The departure from the exact value of T is given by the ratio

$$\frac{T_{\text{oc}}}{T} = 1 + \frac{Z_2}{Z_1} \tag{4.5}$$

The smaller the ratio of Z_2 to Z_1, the more accurate this method becomes in the absence of a proper termination.

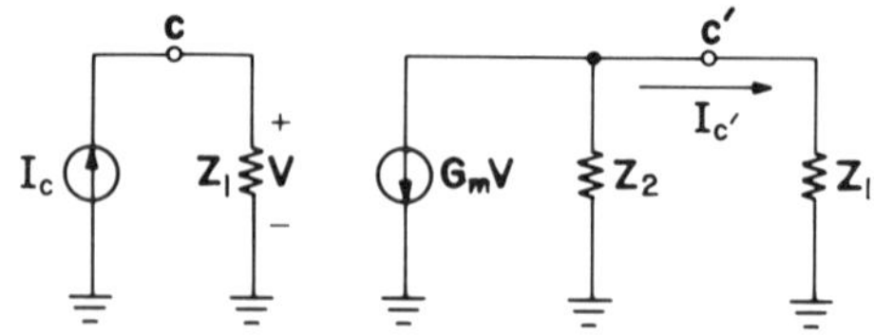

Figure 4.3 Amplifier with feedback loop cut and terminated.

4.2 *Breaking Loop—Current Ratio Measurement*

We start again from Fig. 4.1. The loop is broken at c–c', and the cut end at c' is terminated with the impedance Z_1 as shown in Fig. 4.3. A current source now excites the amplifier on the left and the current $I_{c'}$ is measured on the right.

The ratio of $I_{c'}/I_c$ is calculated and it is found that

$$\frac{I_{c'}}{I_c} = -G_m \frac{Z_1 Z_2}{Z_1 + Z_2} \tag{4.6}$$

We see that we can measure the return ratio by cutting and terminating the feedback loop and calculating the return ratio using

$$T = -\frac{I_{c'}}{I_c} \tag{4.7}$$

As was mentioned in connection with the voltage ratio measurement, we have to terminate at the cut with the impedance that the signal would encounter if it continued around the loop. If the cut is chosen so that the impedances on both sides of the cut are such that Z_1 is small compared to Z_2, then Z_1 can be replaced by a short circuit, but an inaccuracy will be introduced into the measurement. To find the error in the measurement, we observe that when T is calculated using (4.7) and (4.6), and Z_1 is replaced by a short circuit, the approximate return ratio using the current ratio is

$$T_{sc} = G_m Z_1 \tag{4.8}$$

The departure from the exact value of T is very readily expressed by the ratio

$$\frac{T_{sc}}{T} = 1 + \frac{Z_1}{Z_2} \tag{4.9}$$

The smaller the ratio of Z_1 to Z_2, the more accurate this method becomes in the absence of a proper termination.

4.3 *Combining Voltage and Current Gain Methods*

If it is impossible to cut the feedback amplifier loop at a place where there is a substantial impedance mismatch, or the impedance needed to terminate the cut is difficult to ascertain, accurate loop-gain measurements can be performed nonetheless. The feedback loop is cut at any convenient location and a voltage ratio measurement is taken with the cut terminated in an open circuit. The location of the cut remains unchanged but now a current ratio measurement is taken with the cut terminated in a short circuit. We see from (4.4), (4.8), and (4.1) that the actual return ratio T can be calculated from T_{oc} and T_{sc} by using

$$T = \frac{T_{oc}T_{sc}}{T_{oc} + T_{sc}} \tag{4.10}$$

The above can also be written in the form

$$\frac{1}{T} = \frac{1}{T_{oc}} + \frac{1}{T_{sc}} \tag{4.11}$$

From this expression it is clear that the smaller return ratio on the right-hand side of (4.11) controls the accuracy of the return ratio T on the left. If, for example, T_{sc} and T_{oc} can be measured with an accuracy of $\pm 5\%$, and the values are 1 and 20, respectively, then it is clear that there is hardly any point in measuring T_{oc}, since its contribution to the final answer is of the same order of magnitude as the uncertainty in the final answer. It is therefore desirable to exercise greater control over the accuracy of the measurement of the smaller of the two return ratios, which is T_{sc} in this example.

Example 4.1: For the amplifier shown in Fig. 4.4, the loop will be cut at c–c' and the return ratio will be calculated. All the methods discussed so far will be employed.

The calculation of return ratio for this amplifier was carried out in Example 2.1 and it was found that $T = 34.9$. In preparation for measuring the loop gain, the loop is broken at c–c' and the cut is terminated by

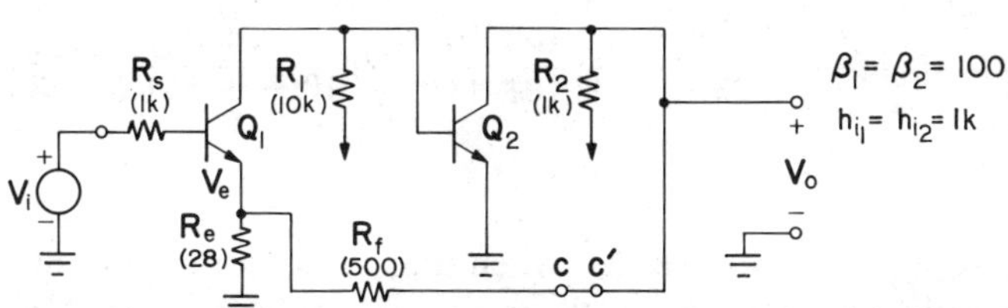

Figure 4.4 Two-transistor feedback amplifier.

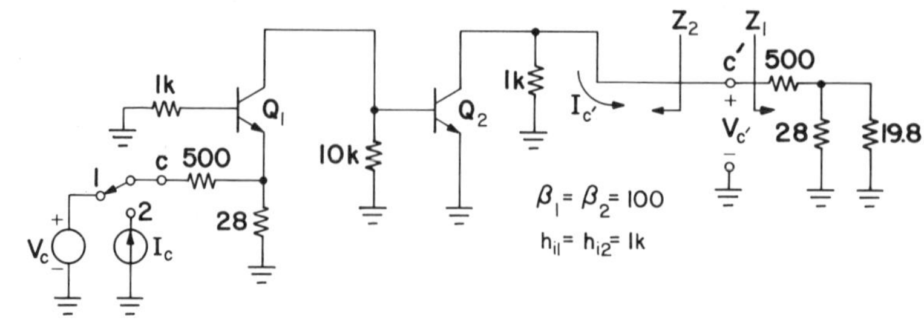

Figure 4.5 Amplifier of Fig. 4.4 prepared for loop-gain measurement.

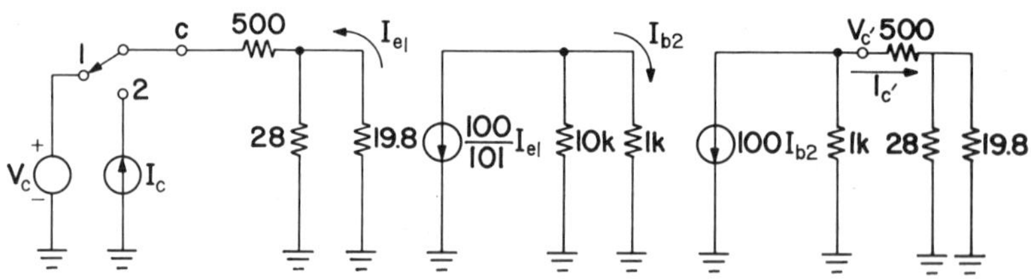

Figure 4.6 Equivalent circuit of Fig. 4.5.

the impedance the signal would encounter if it proceeded around the unbroken loop. The 19.8 Ω shown on the right represents the equivalent resistance seen looking up at the emitter of Q_1. The circuit for measuring loop gain is shown in Fig. 4.5. The equivalent circuit for Fig. 4.5 is shown in Fig. 4.6.

When the switch at the input source is at position 1, we can calculate the return ratio by using (4.3). The calculation is

$$T = -\frac{V_{c'}}{V_c} = \frac{1}{511.6}\,\frac{28}{28 + 19.8}\,\frac{100}{101}\,\frac{10k}{11k}\,100(1k\|511.6) = 34.9 \tag{4.12}$$

The switch at the input is now thrown to position 2 and the calculation of return ratio is carried out according to (4.7). The calculation is

$$T = -\frac{I_{c'}}{I_c} = \frac{28}{28 + 19.8}\,\frac{100}{101}\,\frac{10k}{11k}\,100\,\frac{1000}{1000 + 511.6} = 34.9 \tag{4.13}$$

It comes as no surprise that when the cut is properly terminated, the return ratio agrees with the theoretically obtained value. Now we are ready to calculate the results for an improperly terminated loop.

With the switch in position 1 and the impedance to the right of c' replaced by an open circuit, we calculate T_{oc}:

$$T_{oc} = -\frac{V_{c'}}{V_c} = \frac{1}{511.6}\frac{28}{28 + 19.8}\frac{100}{101}\frac{10k}{11k}100 \cdot 1k = 103.1 \tag{4.14}$$

The ratio of T_{oc} to the actual T is

$$\frac{T_{oc}}{T} = 2.955 \tag{4.15}$$

This large error is due to the fact that $Z_2 \ll Z_1$ does not hold in this case. From (4.5) we see that the above error is due to

$$1 + \frac{Z_2}{Z_1} = 1 + \frac{1k}{511.6} = 2.955 \tag{4.16}$$

With the switch at position 2 and the impedance to the right of c' replaced by a short circuit, we calculate T_{sc}:

$$T_{sc} = -\frac{I_{c'}}{I_c} = \frac{28}{28 + 19.8}\frac{100}{101}\frac{10k}{11k}100 = 52.7 \tag{4.17}$$

The ratio of T_{sc} to the actual T is

$$\frac{T_{sc}}{T} = 1.512 \tag{4.18}$$

This large error is due to the fact that $Z_1 \ll Z_2$ does not hold in this case. From (4.9) we see that the above error is due to

$$1 + \frac{Z_1}{Z_2} = 1 + \frac{511.6}{1k} = 1.512 \tag{4.19}$$

Substituting the results of (4.14) and (4.17) into (4.10) we obtain the exact result

$$T = \frac{T_{oc}T_{sc}}{T_{oc} + T_{sc}} = \frac{103.1 \cdot 52.7}{103.1 + 52.7} = 34.9 \tag{4.20}$$

This agrees with the theoretical value calculated in Example 2.1. ■

4.4 *Iterating the Amplifier to Terminate at the Cut*

To perform the measurements using only the voltage-gain method or the current-gain method, we have to cut the feedback loop and terminate the cut properly. The impedance needed to terminate the cut may vary with

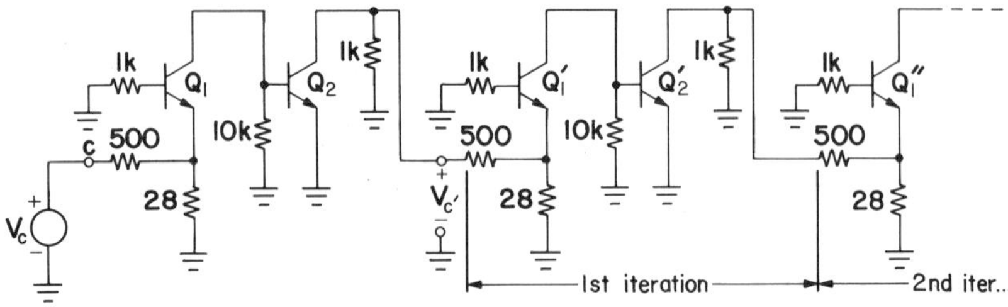

Figure 4.7 Feedback loop cut and terminated by using iteration.

frequency and its frequency dependence may be difficult to determine. To avoid this problem, the cut can simply be terminated in another stage of amplification. The stage would have to correspond to the one that the signal would see next as it traverses the feedback loop. If any uncertainty exists as to whether one additional stage of amplification is adequate to properly terminate the loop, any number of stages can be added to iterate the structure. The iteration process can continue until there is confidence that nothing is to be gained by iterating further. Usually one stage is enough to terminate a feedback loop.

Example 4.2: To illustrate the method, the amplifier of Fig. 4.4 is shown in Fig. 4.7, with the loop cut at c–c', and the cut terminated by using iteration. The cut is shown terminated by repeating three stages of the amplifier. Usually one stage of iteration is adequate, and the three stages of amplification in the iteration in Fig. 4.7 are shown for purposes of demonstrating the method. The negative of the ratio of $V_{c'}$ to V_c determines the return ratio. ■

4.5 *Breaking the Loop Using an Inductor and Terminating*

Suppose the loop cannot be cut because the amplifier's dc-biasing scheme depends on keeping the loop intact for dc signals. This issue was addressed by Rosenstark in [1]. To get around this problem we cut the loop at c–c' in Fig. 4.1 and terminate the right-hand side of the cut in Z_1 in series with a blocking capacitor. We also introduce a large inductor between terminals c and c' as shown in Fig. 4.8. If the inductor's reactance is large enough in the frequency range of interest to prevent ac-signal propagation around the loop, then the loop is effectively cut at c–c'. We are now in a position to measure loop gain according to (4.3) by adding a voltage source V_c and terminating at c' with Z_1.

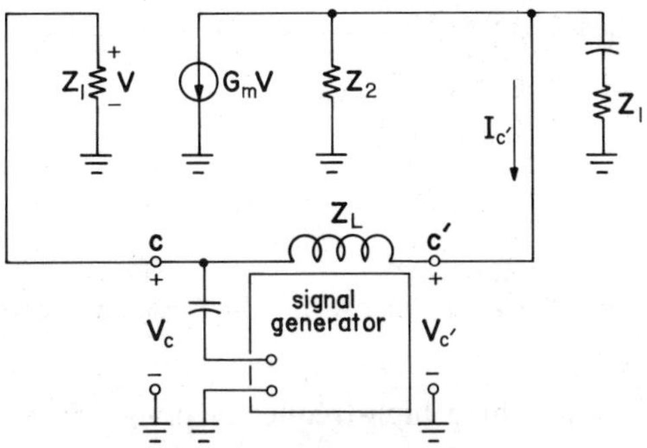

Figure 4.8 Amplifier with feedback loop broken inductively.

The approximate return ratio T_a^L can be calculated from Fig. 4.8 directly

$$T_a^L = -\frac{V_{c'}}{V_c} = G_m(Z_2 \| Z_1 \| Z_L) - \frac{Z_1 \| Z_2}{Z_L + Z_1 \| Z_2} \tag{4.21}$$

The above expression can be rewritten in the form

$$T_a^L = G_m \frac{Z_1 Z_2}{Z_1 + Z_2} \frac{Z_L - 1/G_m}{Z_L + Z_1 \| Z_2} \tag{4.22}$$

We recognize from (4.1) that the first term on the right of the above expression is the actual return ratio T, so we rewrite (4.22) in the form

$$\frac{T_a^L}{T} = \frac{Z_L - 1/G_m}{Z_L + Z_1 \| Z_2} \tag{4.23}$$

It is clear from (4.23) that the ratio T_a^L/T tends to unity if Z_L is chosen so that

$$|Z_L| \gg |1/G_m| \tag{4.24}$$

and

$$|Z_L| \gg |Z_1 \| Z_2| \tag{4.25}$$

so that this method is capable of producing accurate results if the above inequalities are maintained. An illustrative example will now be given to show that this method can be used to produce accurate results.

Example 4.3: The model shown in Fig. 4.9 represents a three-transistor feedback amplifier.

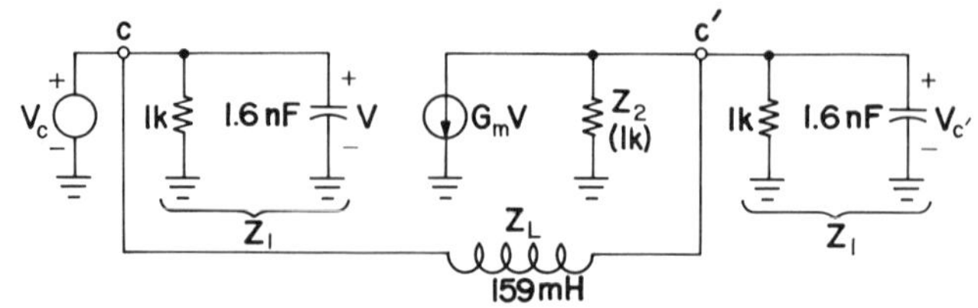

Figure 4.9 Feedback amplifier with inductively cut feedback loop.

The amplifier has three high-frequency poles. Two of the poles are shown incorporated into the expression for the transconductance, which is

$$G_m = \frac{0.5}{(1 + jf/50)^2} \tag{4.26}$$

The variable f is in units of kHz. The third pole results from the frequency-dependent input impedance of the first stage

$$Z_1 = \frac{1000}{1 + jf/100} \tag{4.27}$$

We are interested in measuring the loop gain starting at midband and continuing beyond the loop-gain crossover frequency (which is the frequency at which the magnitude of the return ratio is unity). We shall take midband to be 5 kHz, which is 1 decade below the first critical frequency of 50 kHz. At 5 kHz an inductance of 159 mH for Z_L results in a reactance of 5 kΩ, which is 10 times the parallel combination of Z_1 and Z_2 (which is 500 Ω), so that (4.25) is reasonably satisfied.

At midband (4.24) is also satisfied, but to see if it will still be satisfied at loop-gain crossover, we use the expression for the actual return ratio T to calculate the loop-gain crossover frequency. That expression is found from (4.1)

$$T = G_m \frac{Z_1 Z_2}{Z_1 + Z_2} = \frac{250}{(1 + jf/50)^2(1 + jf/200)} \tag{4.28}$$

By trial and error, it is determined that loop-gain crossover occurs at 486 kHz. At this frequency $|Z_L|$ is 486 kΩ whereas $|1/G_m|$ is 191 Ω, so at loop-gain crossover (4.24) is still well satisfied.

To show the errors involved in carrying out a loop-gain measurement by cutting the feedback loop and inserting an inductance, the ratio of approximate loop-gain T_a^L to actual loop-gain T is calculated using (4.23). The results are tabulated in Table 4.1. It is seen that as frequency increases the errors in measurement fall below the accuracy of measure-

Table 4.1 Errors in magnitude and phase measurement for Example 4.3

f	$\lvert T_a^L/T \rvert$	$\angle T_a^L/T$
5	0.9999	5.75
10	1.0037	2.86
20	1.0045	1.39
50	1.0039	0.46
100	1.0024	0.14
200	1.0009	0.02
390	1.0002	0.01
500	1.0001	−0.02
kHz	—	deg

ment that can be expected in the laboratory. The frequency region in which it is desirable to determine loop gain accurately is the one in the vicinity of gain crossover (which is 486 kHz in this case). It is apparent from Table 4.1 that this objective is readily achieved. ■

We are still left with the problem of measuring the loop gain for feedback amplifiers for which the loop must be kept intact and which have the additional problem that it is difficult to ascertain the impedance needed to terminate the loop. The following methods address this problem. The first method is based on the method of this section.

4.6 *Node Injection—Voltage Ratio*

To make sure that the dc performance of the loop is not affected, a signal is capacitively coupled into the loop at node c as shown in Fig. 4.10. The signal propagation in the feedback loop is blocked by means of an inductor,

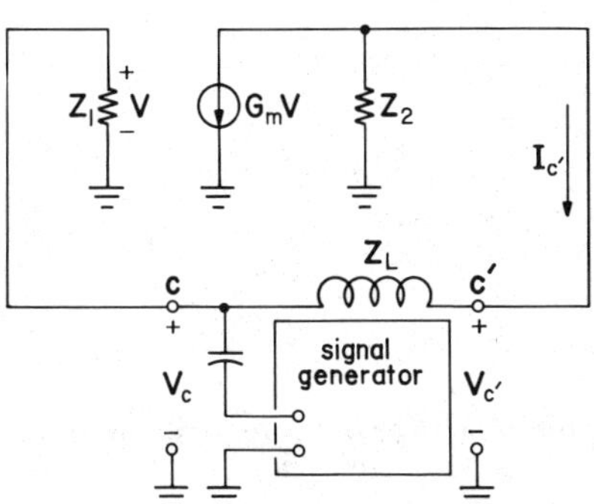

Figure 4.10 Signal injected at a node. Feedback loop signal blocked by an inductor.

whose impedance Z_L is chosen to satisfy

$$|Z_L| \gg |Z_2| \tag{4.29}$$

As a consequence of (4.29) the current $I_{c'}$ is zero. We now evaluate the open circuit return ratio by using

$$T_{oc}^L = -\frac{V_{c'}}{V_c} \tag{4.30}$$

to find

$$T_{oc}^L = G_m Z_2 \tag{4.31}$$

We note in passing that this would be the loop gain of the amplifier if the measurement were performed at a location at which

$$|Z_1| \gg |Z_2| \tag{4.32}$$

We see that this method used by itself has the potential for producing accurate results for special cases.

4.7 *Branch Injection—Current Ratio*

To make sure that the dc performance of the loop is not affected, a signal is transformer coupled into the loop in series with branch c–c' as shown in Fig. 4.11. The signal propagation in the feedback loop is prevented by means of a capacitor, whose impedance Z_C is chosen to satisfy

$$|Z_C| \ll |Z_2| \tag{4.33}$$

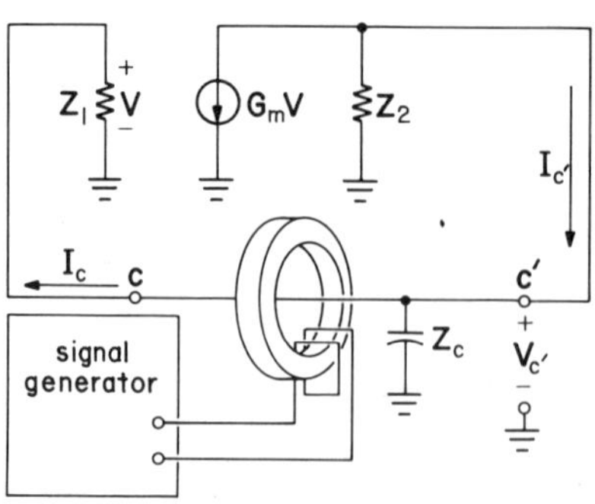

Figure 4.11 Signal injected into a branch. Loop signal shunted to ground by a capacitor.

As a consequence of (4.33), the voltage $V_{c'}$ is zero. We now evaluate the short-circuit return ratio by using

$$T_{\text{sc}}^{C} = -\frac{I_{c'}}{I_c} \tag{4.34}$$

to find

$$T_{\text{sc}}^{C} = G_m Z_1 \tag{4.35}$$

In this case, we note that this would be the loop gain of the amplifier if the measurement were made at a location at which

$$|Z_1| \ll |Z_2| \tag{4.36}$$

We see that this method carried out by itself has the potential for producing accurate results for special cases.

4.8 *The Exact Result—Combining the Measurements*

The method of Sec. 4.6 is useful for measuring loop gain in feedback amplifiers in which a node is accessible at which inequality (4.32) applies. The method of Sec. 4.7 is useful in feedback amplifiers possessing an accessible node at which inequality (4.36) is valid. We need a method that will work for all amplifiers without any constraints. By comparing (4.1), (4.31), and (4.35), we see that we can calculate the loop gain T by using

$$\frac{1}{T} = \frac{1}{T_{\text{oc}}^{L}} + \frac{1}{T_{\text{sc}}^{C}} \tag{4.37}$$

Middlebrook [2, 3] presented a similar method, which also required that two measurements be performed and then an equation similar to (4.37) be used to find the loop gain. The technique—the double null method—required two signals, with a specific amplitude and phase relationship, which had to be used simultaneously in order to carry out the loop-gain measurement. This meant that the two signals had to be derived from the same signal source, with the attendant need for buffer amplifiers, variable attenuators, and variable phase-shift networks. The advantage of the method presented in this section lies in the fact that the two measurements needed to obtain T_{oc}^{L} and T_{sc}^{C}, can be made using relatively simple instrumentation (the most complicated items required are coils with large inductance and some current transformers or current probes). This method dispenses with the need for two ac signals with a specific amplitude and phase relationship. It is useful for all feedback amplifiers without any constraints, but as is the case in Middlebrook's double null method, it requires that the loop gain be

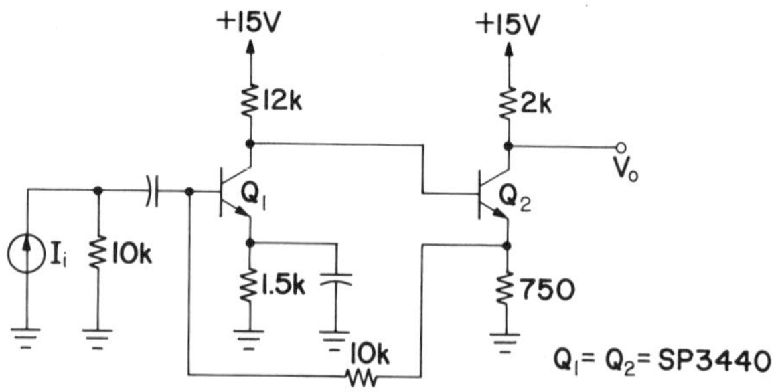

Figure 4.12 Two-transistor feedback amplifier.

measured at every frequency of interest by two methods, in effect doubling the amount of work that has to be performed in the laboratory.

Example 4.4: A Two-Transistor Feedback Amplifier

To get a feel for the method, the two-transistor dc-coupled feedback amplifier of Fig. 4.12 was thoroughly analyzed by use of digital computer circuit analysis programs. It was decided to simulate the measurements on a computer, so that the data would not contain any experimental errors, and also the question of the accurate representation by the model of the transistors selected for the test would not have to be raised.

This amplifier was designed to yield a transimpedance of 28 V/mA. The fact that it is dc coupled presents problems when attempts are made to perform the loop-gain measurement, since physically breaking the loop will upset the dc bias of both transistors. The first transistor has a collector current of 1 mA and the second transistor has a collector current of 3 mA. The hybrid-pi transistor model parameters were calculated from the manufacturer's specifications and the equivalent circuit is shown in Fig. 4.13.

Since we have a substantial impedance mismatch (approximately 92 : 1) for a signal traversing the loop from right to left at nodes b'–b,

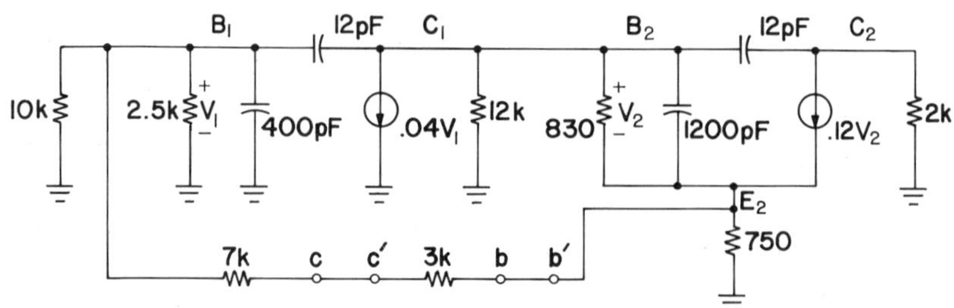

Figure 4.13 Equivalent circuit of the amplifier.

then the node injection method of Sec. 4.6 applied at this point can be expected to yield very accurate results and use of the branch injection method of Sec. 4.7 would be superfluous. When the short circuit b–b' is cut, it is found that the output impedance at node b' is less than 110 Ω. An inductance of 200 mH inserted between terminals b and b' will have a reactance of 1260 Ω at the midband frequency of 1 kHz. Thus the inequality (4.29) is modestly satisfied with a ratio of impedances slightly exceeding 10.

To proceed with the analysis, a voltage source is attached from node b to ground and the voltage is calculated at node b'. The negative of the voltage gain is the return ratio. The results of this calculation are shown in the curves labelled (b) in Fig. 4.14. The curves labelled (a) were obtained for the feedback amplifier loop gain, with the loop cut and properly terminated, something that can be done computationally for a dc-coupled amplifier in spite of the fact that it could not be readily performed in the laboratory. The curves labeled (a) serve as a reference for all the other calculations.

To have a reasonable example for which the use of the method of this section cannot be avoided, we deliberately chose to test the amplifier at the nodes marked c–c'. There is no particular impedance mismatch at this location so we have to obtain T_{oc}^{L} and T_{sc}^{C}, and then calculate the loop gain T using (4.37). A 5-H inductance was chosen to connect nodes c–c'. This inductance has a reactance of over 31 kΩ at the midband frequency of 1 kHz; this value being slightly greater than 10 times the resistance seen to the right of node c'. This size inductance can be readily constructed, in practice, using a toroidal core. If in performing a practical loop-gain measurement, the self-resonant frequency of the coil is reached as the test frequency is increased well above midband, then the coil can be replaced with one having a much smaller inductance, and the test continued, the only consideration being that (4.29) should be satisfied.

To measure T_{sc}^{C}, a capacitor of 100 μF is connected from node c' to ground, and a signal source is connected from node c' to node c. The negative of the ratio of the currents $I_{c'}$ and I_c is T_{sc}^{C}. When this result is combined in (4.35) with the T_{oc}^{L} obtained with the 5-H inductance, we get the curves labelled (c) in Fig. 4.14.

To see what would happen if we have an inadequate impedance mismatch needed to satisfy (4.29), the last calculations were redone, but the 5-H inductance was replaced with a 500-mH inductance. The results of this calculation are shown in the curves labelled (d) in Fig. 4.14.

It is well known that a transfer-function critical point (a transfer-function pole or a zero) has almost no influence on the amplitude characteristic at a frequency one decade below the break frequency, whereas it changes the phase characteristic by a fairly significant 5.7°. A

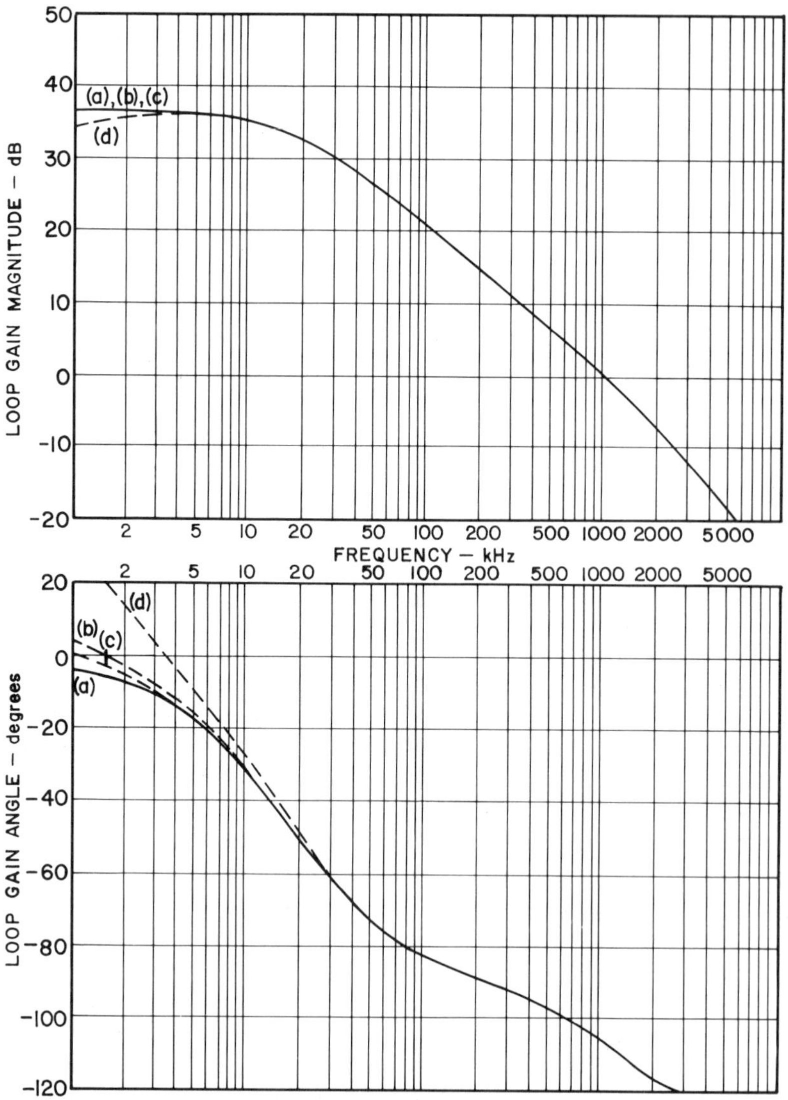

Figure 4.14 Loop gain amplitude and phase characteristics. (*a*) Loop cut and terminated properly. (*b*) T_{oc}^{L} at *b–b′* with 200-mH coil. (*c*) T_{oc}^{L} with 5-H coil and T_{sc}^{C} with 100-μF capacitor at *c–c′*, combined using (4.37). (*d*) Same as (*c*) but with a 500-mH coil.

similar effect can be seen in the results plotted in Fig. 4.14. It is apparent that the phase characteristic is much more sensitive to improper selection of the reactance which is used to open the feedback loop, than is the amplitude characteristic. But even if the ratio of impedances is unity, so that there is no significant impedance mismatch at the point at which the loop is broken, which is the case for the curves labelled (d), the phase characteristic is still very accurate at gain crossover. Of course, we are dealing with a midband loop gain of over 36 dB. For a smaller loop gain this would not hold. It is reasonable to conclude that it is safest to have an impedance mismatch of at least 10 in the range of frequencies at which loop-gain measurements are taken. ■

4.9 *Conclusion*

The methods which were presented in this chapter show that it is possible to perform accurate loop-gain measurements in feedback amplifiers including those in which the feedback loop has to be kept intact for dc signals. It is easiest to measure the loop gain of feedback amplifiers in which it is possible to break the loop and terminate properly. If the loop has to be kept intact for dc signals, then opening the loop using an inductor and terminating properly is a little more complicated but can be carried out without too much difficulty. The most complicated case is that presented by the amplifier in which the loop has to be kept intact because of dc considerations, and for which it is not possible to ascertain the impedance required to properly terminate the loop. In this case the methods of Secs. 4.6–4.8 are a little more trouble to carry out, but the results are well worth the extra effort.

The methods of this chapter are easy to apply because the equipment required consists of readily available instruments. The most exotic devices required are coils of large inductance for the measurement of T_{oc}^L, and current probes and current transformers for the measurement of the currents for determining T_{sc}^C. Finally, if the point at which the measurements are taken has a large impedance mismatch, then one set of measurements is sufficient. If such a point in the feedback amplifier cannot be found, or happens to be inaccessible, then two sets of measurements make it possible to produce accurate results at the expense of a slight inconvenience.

REFERENCES

1. Rosenstark, S.: Loop Gain Measurement in Feedback Amplifiers, *Int. J. of Electronics*, vol. 57, no. 3, pp. 415–421, March, 1984.

2. Middlebrook, R. D.: Measurement of Loop Gain in Feedback Systems, *Int. J. of Electronics*, vol. 38, pp. 485–512, April, 1975.

3. Middlebrook, R. D.: Improved-Accuracy Phase Angle Measurement, *Int. J. of Electronics*, vol. 40, pp. 1–4, January, 1976.

EXERCISES

4.1. The remark was made in Example 4.4 that "it is well known that a transfer-function critical point (a transfer-function pole or a zero) has almost no influence on the amplitude characteristic at a frequency one decade below the break frequency, whereas it changes the phase characteristic by a fairly significant 5.7°." Evaluate the magnitude (in dB) and the phase for the transfer function

$$H(f) = \frac{1}{1 + jf/f_p}$$

at the frequency $f = f_p/10$.

4.2. For the shunt–series two-transistor feedback amplifier shown in Fig. P4.2 it is assumed that the loop is not dc coupled.

(a) If the loop is broken at a–a', draw the excitation (voltage or current source) and the proper termination and determine the loop gain using Eq. (4.3).

(b) How large is the error if the cut is unterminated?

(c) The impedance from $B1$ to ground is frequency dependent and takes some trouble to determine, so it is decided to replace it with a short circuit in the termination. How large is the error in measurement at midband and how does the error change as frequency increases?

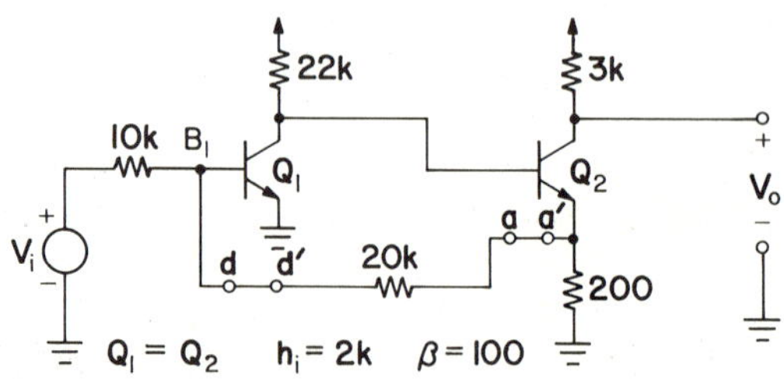

Figure P4.2

4.3. For the shunt–series two-transistor feedback amplifier shown in Fig. P4.2, it is assumed that the loop is not dc coupled.

(a) If the loop is broken at d–d', draw the excitation (voltage or current source) and the proper termination and determine the loop gain using Eq. (4.7).

(b) How large is the error if the cut is terminated in a short circuit?

(c) The impedance from $B1$ to ground is frequency dependent and takes some trouble to determine, so it is decided to replace it with a short circuit in the termination. How large is the error in measurement at midband and how does the error change as frequency increases?

4.4. For the shunt–series two-transistor feedback amplifier shown in Fig. P4.2 it is assumed that the loop is not dc coupled.

(a) If you have not previously solved for the loop gain in Prob. P2.3 then now is a good time to do so.

(b) The loop is opened at a–a' and the cut is terminated in an open circuit. Calculate the loop gain which would be measured by the voltage ratio method.

(c) The loop is opened at a–a' and the cut is terminated in a short circuit. Calculate the loop gain, which would be measured by the current ratio method.

(d) Combine the results of parts (b) and (c) using Eqs. (4.10) or (4.11). The result should correspond to the value calculated in part (a).

4.5. For the shunt–series two-transistor feedback amplifier shown in Fig. P4.2 it is assumed that the loop *is* dc coupled. Assume that midband occurs at 1 kHz.

(a) The loop is cut at a–a' and an inductor is inserted. Draw the model equivalent to Fig. 4.8 and assign numerical values to all the parameters.

(b) Choose a value of inductance which satisfies (4.25) by a factor of 10. Make sure (4.24) is also satisfied.

(c) What will be the error in the loop-gain measurement at midband?

4.6. For the shunt–series two-transistor feedback amplifier shown in Fig. P4.2 it is assumed that the loop *is* dc coupled.

(a) The loop is cut at a–a' and an inductor is inserted. Draw the model equivalent to Fig. 4.10 and assign numerical values to all the parameters.

(b) Choose a value for the inductance L which will satisfy (4.29) by a factor of 10. Will the inequality improve as frequency increases?

(c) Calculate the loop gain T_{oc}^{L} on the assumption that L is infinite, and also T_{a}^{L} for the value of L chosen in part (b). Compare these values to the theoretical loop gain found in Prob. 4.4a.

(d) Draw the model equivalent to Fig. 4.11. Choose a value for the capacitance C which satisfies (4.33) by a factor of 10. Will the inequality improve as frequency increases?

(e) Calculate the loop gain T_{sc}^{C} on the assumption that C is infinite, and also T_{a}^{C} for the value of C chosen in part (d). Compare these values to the theoretical loop gain found in Prob. 4.4a.

(f) Substitute the results of parts (c) and (e) into (4.37).

5

High-Frequency Analysis of Loop Gain

WE HAVE SO FAR DEALT with the problems of calculating the feedback amplifier loop gain without any concern that this quantity may be frequency dependent. The solution is not complete until the frequency dependence of the loop gain is determined. Without this knowledge it is impossible to predict the stability of the feedback amplifier (Chap. 6), and the steps needed to obtain a desired frequency response or transient response (Chap. 7).

Various methods are used to find the frequency response of electronic circuits. The most common method is to determine the poles of the frequency response using the Miller approximation (or the Miller effect). It will be demonstrated that this method gives reasonably good results only for the dominant loop-gain pole, and is therefore relatively useless for obtaining frequency response data at frequencies beyond the first pole, which is to say that it does not give good results beyond the loop gain 3-dB frequency. It will be seen in Chap. 6, that to assess the stability of feedback amplifiers, it is necessary to know the loop-gain frequency response at loop-gain crossover. This is the frequency at which the loop gain equals unity (or 0 dB), and this frequency usually lies substantially beyond the loop-gain 3-dB frequency. In spite of the shortcomings of the Miller approximation, it will nonetheless be covered in Sec. 5.4, and it is applied to the examples in this chapter for purposes of comparison, because it can at the very least be used to obtain a very crude idea of the high-frequency behavior of the loop gain.

In addition to the Miller approximation, which is of limited usefulness for feedback amplifier loop-gain calculations, methods will be presented which can be used for finding transfer function critical points by using hand calculation. Even when powerful computer circuit analysis programs such as PCAP, CORNAP, and SPICE are available, a great deal of insight may still be gained by use of these methods of hand calculation. The above techniques will be presented in later sections. The first item of interest will be a review of useful equivalent circuits for performing high-frequency response calculations.

5.1 *High-Frequency Models for Bipolar Transistors and FETs*

When transistors first came into use, there were many different models available for the calculation of the high-frequency response of electronic circuits. Eventually the hybrid-pi model became accepted as the most reasonable for performing frequency response calculations. The advantage of this model is that it is equally useful for bipolar transistors and FETs. The transistor and FET models, which are presented in App. A and used in the previous chapters, are useful for performing low-frequency response calculations, and they are related to the hybrid-pi model which will be discussed presently. This presentation is given as a review. To get a very thorough discussion of bipolar-transistor and FET models, the reader should consult Ref. [1].

To determine the parameters of the hybrid-pi bipolar-transistor model shown in Fig. 5.1 it is necessary to know the collector bias current of the transistor. For a collector bias current of I_C we can calculate g_m at room temperature using

$$g_m = 40 I_C \tag{5.1}$$

The transistor current gain β is found in the transistor manufacturer's specification sheet. With this information and the value of g_m calculated from (5.1), r_π can be calculated using the relationship

$$\beta = g_m r_\pi \tag{5.2}$$

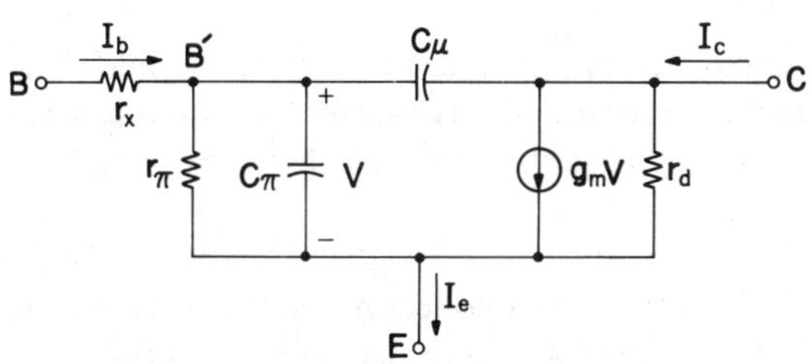

Figure 5.1 Hybrid-pi equivalent circuit.

Other items contained in the manufacturer's data are values for the current-gain · bandwidth product f_T and the output capacitance C_μ, sometimes identified as C_{ob} and also as C_{cb}. The value of C_π can be calculated from the relationship

$$f_T = \frac{g_m}{2\pi(C_\pi + C_\mu)} \tag{5.3}$$

The ohmic (base-spreading) resistance r_x is usually not specified, and may be assumed to have a value between 10 and 500 Ω. If it is disregarded, then the effect on the frequency response is usually not very great. In the absence of a specified value an assumption of 100 Ω is quite reasonable for a low-frequency transistor. The transistor output resistance r_d is usually not found in the manufacturer's data. It is some multiple of 10 kΩ in most transistors. Since most transistor circuits contain a collector resistance of a few kiloohms or less, then ignoring the output resistance will have a small effect on frequency response calculations, and that is usually what is done.

The same model that was used for bipolar transistors can also be used for FETs, but r_x must be replaced by a short-circuit, and r_π must be replaced by an open circuit. The values of g_m and r_d can be found directly in the manufacturer's specifications. The low-frequency model parameter μ, if desired, can be found from the relationship

$$\mu = r_d g_m \tag{5.4}$$

The capacitance C_μ is found in the manufacturer's specifications as C_{gd}, C_{rs}, or C_{rss}. C_π is sometimes specified directly as C_{gs}. Otherwise, if C_{is} or C_{iss} is specified, then C_π can be found from the relationship

$$C_{is} = C_{iss} = C_\pi + C_\mu \tag{5.5}$$

The above is given as a review. It must be understood that almost all the parameters discussed above are dependent on the point of device operation and temperature, and the manufacturer's instructions on the use of the specification sheets should be consulted carefully.

5.2 *The Complete Loop-Gain Solution*

To be able to make a comparison of the various methods for calculating the loop-gain frequency response, the two-transistor amplifier example shown in Fig. 5.2 was very thoroughly analyzed by use of the computer circuit analysis program CORNAP.

This dc coupled amplifier was designed to have a transimpedance of 28 V/mA. Both transistors have a collector current of 1.33 mA. The hybrid-pi model transistor parameters were calculated from the manufacturer's specifications and the equivalent circuit is shown in Fig. 5.3.

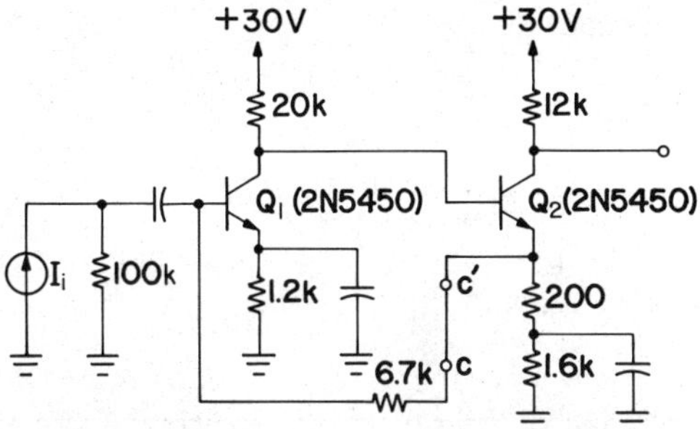

Figure 5.2 Two-transistor feedback amplifier.

The loop is cut at c–c' in accordance with the principles of Chap. 4 and a voltage source is connected from terminal c to ground. The output impedance at terminal c' is less than 105 Ω, whereas the input impedance at terminal c lies between 8660 Ω at midband and 6800 Ω at very high frequencies. The cut in the loop was terminated with a 6800-Ω resistance, so that the error in loop-gain measurement would be minimized at high frequencies, where it is desired to have greater accuracy. Since the loop gain depends on the parallel combination of the output and input impedance at the cut, as indicated by (4.1), then the error in loop-gain measurement due to improper termination at low frequency cannot be larger than 0.33%.

The poles and zeros for the loop gain were determined by using the CORNAP circuit analysis program, and it was found that the poles are located at

$$\begin{aligned} p_1 &= -0.139 \text{ Mrad/s} \\ p_2 &= -9.57 \text{ Mrad/s} \\ p_3 &= -682 \text{ Mrad/s} \\ p_4 &= -1353 \text{ Mrad/s} \end{aligned} \tag{5.6}$$

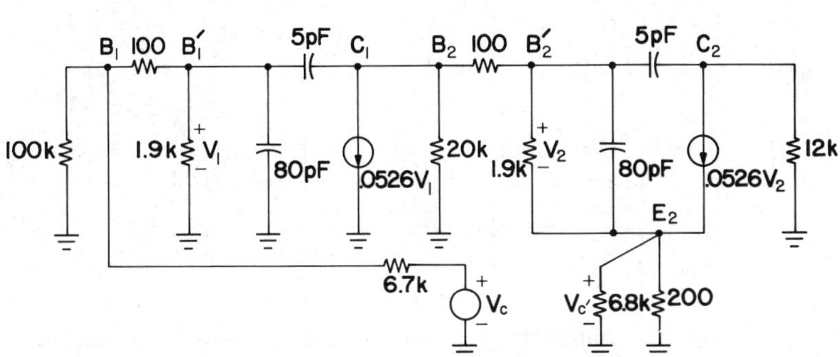

Figure 5.3 Two-transistor amplifier equivalent circuit.

and the zeros are located at

$$\begin{aligned} z_1 &= -16.7 \text{ Mrad/s} \\ z_2 &= -665 \text{ Mrad/s} \\ z_3 &= +10526 \text{ Mrad/s} \end{aligned} \tag{5.7}$$

The loop gain corresponding to these critical points is shown in Fig. 5.4, curve *a*. Clearly poles p_3 and p_4 and zeros z_2 and z_3 are so far removed

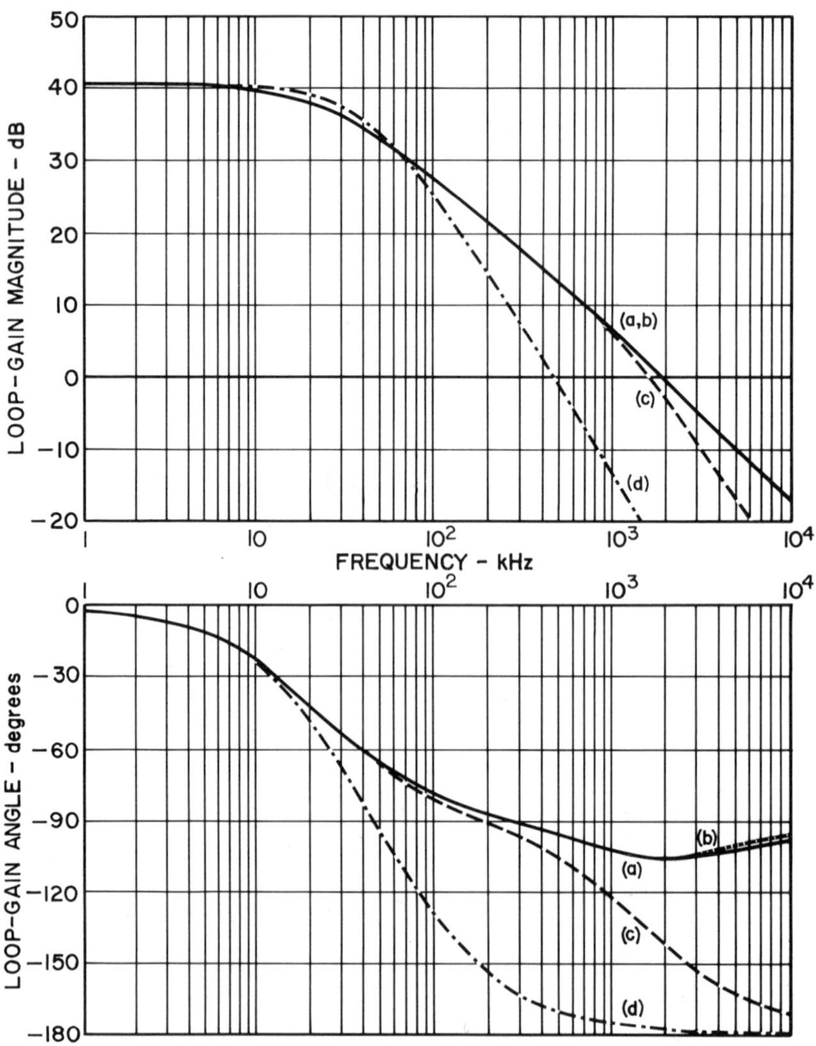

Figure 5.4 Loop-gain amplitude and phase characteristics. (*a*) Loop gain with all poles and zeros. (*b*) Loop gain with only significant poles and zeros. (*c*) Loop gain with significant poles only. (*d*) Loop gain with poles calculated using the Miller effect.

from the origin that they should have a negligible effect on the loop-gain characteristic. Curve *b* was plotted to see what would happen if the poles and zeros at very high frequencies are totally ignored. We see that these results depart so negligibly from those of curve *a*, that the amplifier loop-gain characteristic can be very adequately obtained from what will be henceforth referred to as the *significant* or *principal* poles and zeros, in this case p_1, p_2, and z_1.

The question that comes to mind now is whether a simple method exists for finding the principal poles and zeros without resorting to computer calculation.

5.3 *The Zeros of Transistor Circuits*

For the purpose of deriving equations for the calculation of the zeros of transistor circuits, the source-free hybrid-pi model of Fig. 5.5 will be analyzed. This model describes bipolar and field effect transistors equally well. The active device is shown resistively loaded, because the assumption at this point is that the other active-device stages are not yet exhibiting frequency-dependent behavior. In a bipolar transistor the terminal marked 1 would correspond to B', hence R_i would have to include the effects of r_x, the ohmic resistance connecting the base terminal B to B'. In an FET circuit, terminal 1 would correspond to the gate and the conductance g_π would be zero. The correspondence of the other terminals is self-explanatory.

The nodal admittance matrix can be written by inspection from Fig. 5.5 in terms of the Laplace transform variable s,

$$Y = \begin{bmatrix} G_i + g_\pi + s(C_\pi + C_\mu) & -(g_\pi + SC_\pi) & -SC_\mu \\ -(g_m + g_\pi + sC_\pi) & (G_e + g_m + g_\pi + sC_\pi) & 0 \\ g_m - sC_\mu & -g_m & (G_L + sC_\mu) \end{bmatrix} \tag{5.8}$$

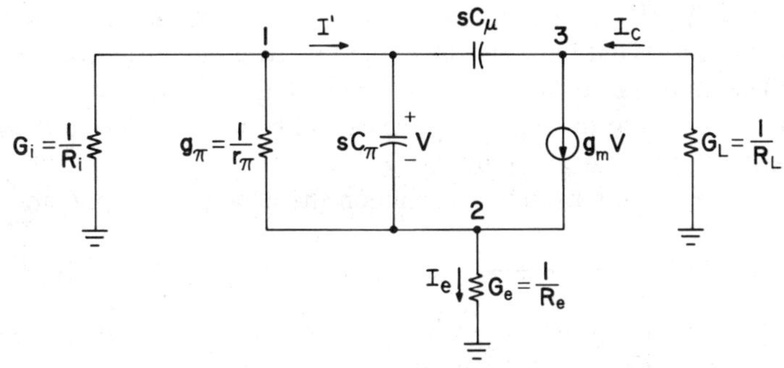

Figure 5.5 Source-free model for deriving the high-frequency behavior of a single transistor stage.

To abbreviate the notation we define

$$g_t \equiv g_m + g_\pi = g_m(1 + 1/\beta) \approx g_m \tag{5.9}$$

The numerator polynomials are determined from the routing of the signal. For signal transmission from base to collector (or gate to drain), the numerator polynomial is

$$\Delta_{13} = R_e C_\pi C_\mu s^2 + (1 + g_t R_e) C_\mu s - g_m \tag{5.10}$$

For signal transmission from base to emitter (or gate to source), the numerator polynomial is

$$\Delta_{12} = (R_L C_\mu s + 1)(C_\pi s + g_t) \tag{5.11}$$

For signal transmission from emitter to collector (or source to drain), the numerator polynomial is

$$\Delta_{23} = R_i C_\pi C_\mu s^2 + R_i g_t C_\mu s + g_m \tag{5.12}$$

It is noteworthy that (5.10) shows that right-half plane zeros are possible in active devices, but we have already seen in the two-transistor example, that the right-half plane zero z_3 is so far removed from the origin, as to have no significant effect on the transfer function. Even if the right-half plane zero did have a significant effect on the transfer function, it would still be no reason for any misgivings. Networks which are said to be stable (namely those networks which have decaying transient responses), must have poles which lie only in the left half of the s plane, but there is no constraint on where the zeros must lie.

The results obtained from (5.10)–(5.12) can be very accurate. For example, the only noncapacitive impedance appearing in (5.10) is R_e. This equation finds the zeros of transmission for a signal going from base to collector. If the impedance at the *third terminal*, the terminal not directly involved in the propagation of the signal (in this case, the emitter), is connected to a resistor, then the solution due to (5.10) will have no error. If the third terminal is connected to a frequency-dependent impedance, for which it is possible to determine the frequency-dependent expression, then (5.10) will still produce accurate results, but the equation will no longer be of second order. What has been said about (5.10) is also true for (5.11) and (5.12). The impedance at the third terminal is easy to control in most electronic circuits, so that (5.10)–(5.12) can all be used to solve for the zeros of transfer functions without any error.

All transfer functions have in common the denominator polynomial

$$\Delta = C_\pi C_\mu s^2 + \left[\left(\frac{1}{R_L + R_i \| R_e}\right) C_\pi + \left(g_t + \frac{1}{R_e + R_i \| R_L}\right) C_\mu\right] s + \frac{1 + g_t R_e + g_\pi R_i}{R_e(R_L + R_i) + R_L R_i} \tag{5.13}$$

Table 5.1 Summary of parameters from Fig. 5.3 needed for substitution into (5.10)–(5.12)

	Transistor 1	Transistor 2	Units
g_m	0.0526	0.0526	mho
g_π	0.000526	0.000526	mho
g_t	0.0532	0.0532	mho
R_L	10382	12000	ohm
R_i	6380	20100	ohm
R_e	0	194	ohm
C_π	80	80	pF
C_μ	5	5	pF

The use of (5.13) for the finding of the poles for electronic circuits is not recommended, since the assumption that the stage under analysis is resistively loaded at the input and the output does not generally hold. This equation is included here for completeness only and should be used only in situations in which its limitations are clearly understood.

To see what kind of numbers (5.10)–(5.12) produce, the equations were applied to the two-transistor amplifier of Figs. 5.2 and 5.3. Table 5.1 summarizes the quantities derived from Fig. 5.3, which are needed for substitution into those equations.

Substitution for transistor 1 into (5.10) and for transistor 2 into (5.11) results in

$$\Delta_{13} = 5 \times 10^{-12}s - 0.0526 \tag{5.14}$$

and

$$\Delta_{12} = (6 \times 10^{-8}s + 1)(80 \times 10^{-12}s + 0.0532) \tag{5.15}$$

The resultant zeros are

$$\begin{aligned} z_1' &= -16.7 \text{ Mrad/s} \\ z_2' &= -665 \text{ Mrad/s} \\ z_3' &= +10526 \text{ Mrad/s} \end{aligned} \tag{5.16}$$

A comparison of (5.16) and (5.7) shows that there is no difference between the results which were obtained by using CORNAP and those obtained by use of (5.10)–(5.12).

Now that we have a method for hand-calculating the zeros, we need a corresponding method for finding the loop-gain poles. The desired method should be suitable for hand calculation, and should not be too difficult to apply when only principal poles are desired.

5.4 *The Poles of Transistor Circuits—the Miller Effect*

One of the best known techniques for finding the poles of electronic circuits is the Miller approximation, more commonly referred to as the Miller effect [2]. The effect was used to explain why electronic devices (at that time, vacuum tubes) with seemingly low input capacitance, behaved at high frequencies as if the input capacitance was much larger than measured. The method used for explaining this behavior became one of the most popular techniques for calculating the high-frequency response of electronic amplifiers. It has been mentioned previously that this method is useful for finding the lowest pole, namely the 3-dB frequency of an amplifier. The range of frequencies of interest to the designer of feedback amplifiers is a band in the vicinity of loop-gain crossover, and in this region the Miller approximation yields very poor results. But in the absence of other, more convenient techniques, the Miller approximation can be used to get a rough estimate of the frequency behavior of electronic circuits, and for that reason it will be reviewed presently.

The analysis will proceed from Fig. 5.5. All the remarks made at the beginning of Sec. 5.3 are still appropriate. The single stage shown has two energy storage devices, the capacitors C_π and C_μ. To avoid having to deal with the second-order denominator polynomial (5.13) to which this network gives rise, it is desired to remove the capacitor C_μ, and replace C_π with a capacitor of a different value C_π', so that the current I' will remain unchanged at midband. The voltage to the left of C_μ is

$$V_{12} = V \tag{5.17}$$

Midband is the range of frequencies that is sufficiently low so that all circuit capacitors can be considered to be open circuits. To find the voltage V_{32} at midband we observe that the current in R_L is

$$I_c = g_m V \tag{5.18}$$

whereas the current in the resistor R_e is

$$I_e = \frac{V}{r_\pi} + g_m V \tag{5.19}$$

Using the above we can calculate

$$V_{32} = -g_m V R_L - \left[\frac{V}{r_\pi} + g_m V\right] R_e \tag{5.20}$$

The voltage across C_μ is given by the difference between (5.17) and (5.20)

$$V_{13} = \left[1 + g_m R_L + \left(\frac{1}{r_\pi} + g_m\right) R_e\right] V \tag{5.21}$$

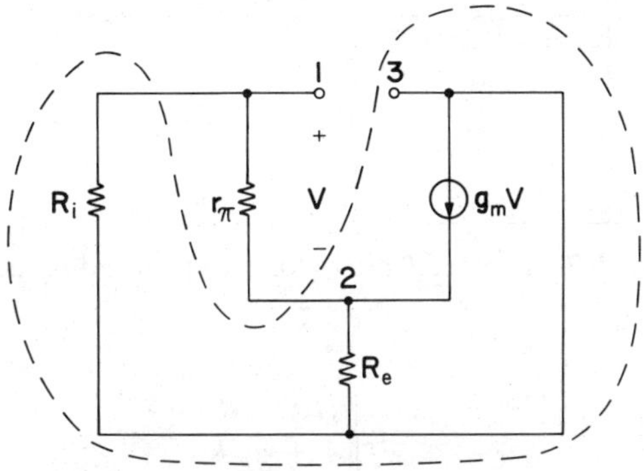

Figure 5.6 Figure 5.5 without capacitors.

From (5.2) we see that g_m is greater than $1/r_\pi$ by a factor β, so that the above expression can be simplified somewhat into

$$V_{13} \approx [1 + g_m(R_L + R_e)]V \tag{5.22}$$

The current in C_μ is the above voltage multiplied by the admittance sC_μ. The total current I' is therefore given by

$$I' \approx \{sC_\pi + [1 + g_m(R_L + R_e)]sC_\mu\}V \tag{5.23}$$

It is clear that in the vicinity of midband I' will remain unaffected if we omit C_μ and replace C_π with a larger capacitor C_π' whose value is

$$C_\pi' \approx C_\pi + [1 + g_m(R_L + R_e)]C_\mu \tag{5.24}$$

We can therefore use the above reduction to go from a second-order network to a first-order network. Once that is accomplished, we need to determine the resistance r_π' which shunts the capacitor C_π' across the terminals 1 and 2. It is easiest to start with a fresh diagram in which there are no capacitors as seen in Fig. 5.6.

The resistor R_L was in series with the current source g_mV, so it is not shown in this figure, because it has no effect on the value of the resistance seen between terminals 1 and 2.

By replacing the network inside the dashed lines with a Norton equivalent, we obtain Fig. 5.7.

Since the current source on the left is dependent on the voltage V, which is across it, it can now be replaced by an equivalent resistance whose value is shown in Fig. 5.8*a*. Combining resistors in parallel leads to Fig. 5.8*b*. We see that the resistance between terminals 1 and 2, which represents r_π' is

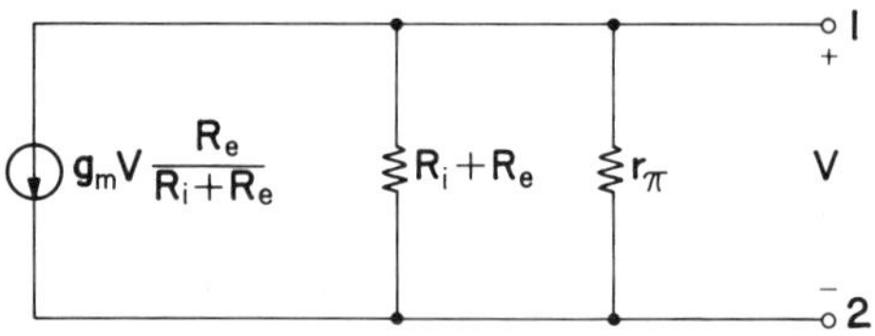

Figure 5.7 Figure 5.6 after Norton equivalent reduction.

given by

$$r'_\pi = r_\pi \left\| \frac{R_i + R_e}{1 + g_m R_e} \right. \tag{5.25}$$

Using the results of (5.24) and (5.25), we find the pole of this resistor–capacitor combination

$$f_p = \frac{1}{2\pi r'_\pi C'_\pi} \tag{5.26}$$

We run into some difficulty with (5.24) if R_e approaches infinity, a situation that would apply to the upper transistor in a cascode connection (see Prob. 5.2). To resolve this dilemma we go back to (5.13) and allow $R_e \to \infty$. The result is

$$\Delta|_{R_e \to \infty} = C_\pi C_\mu s^2 + \left[\frac{C_\pi}{R_L + R_i} + g_t C_\mu \right] s + \frac{g_t}{R_L + R_i} \tag{5.27}$$

It will be shown later that transistors or FETs contribute a significant (low-frequency) pole to the transfer function, and a pole which is at a very much higher frequency than the significant pole. We are interested in the significant pole only, since the Miller approximation is not valid at much higher frequencies. Since the roots of (5.27) are widely separated, we can find the lower root by solving the reduced version of the quadratic. Namely

$$\left[\frac{C_\pi}{R_L + R_i} + g_t C_\mu \right] s + \frac{g_t}{R_L + R_i} = 0 \tag{5.28}$$

to find that for this case the pole is given by

$$f_p|_{R_e \to \infty} = \frac{g_t}{2\pi \left[C_\pi + g_t (R_L + R_i) C_\mu \right]} \tag{5.29}$$

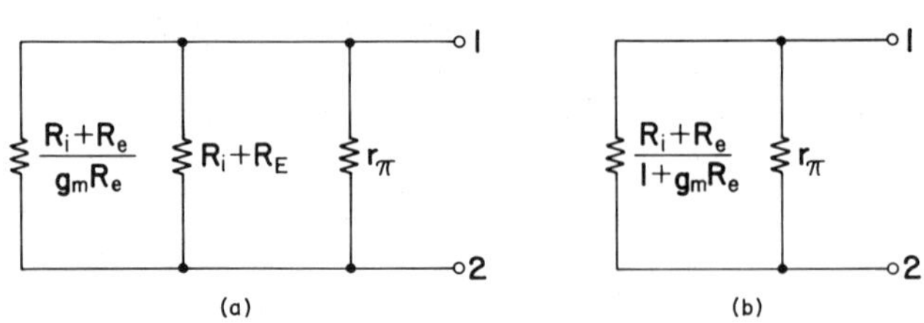

Figure 5.8 (*a*) Figure 5.7 after current source replacement. (*b*) Final reduction.

Table 5.2 Miller approximation results for the two-transistor amplifier of Fig. 5.3

	Transistor 1	Transistor 2	Units
C_π'	2815	3292	pF
r_π'	1464	927.3	ohm
f_p	38.6	52.1	kHz

To illustrate the method, the Miller approximation was applied to the data of Table 5.1, which lists the parameters of the two-transistor amplifier of Fig. 5.3. Table 5.2 summarizes the results obtained.

These results are not in the best agreement with those of (5.6), which were obtained by computer. There the first two poles were found to be at 22.1 kHz and 1.52 MHz. It is not at all surprising that the curve obtained using the Miller effect, which appears in Fig. 5.4, is in such poor agreement with the curves obtained by more accurate methods.

5.5 *The Poles of Transistor Circuits—the Cochrun–Grabel Method*

We are now interested in finding the poles of active *RC* circuits more accurately than the Miller approximation allows. The method of Cochrun and Grabel [3], which was elaborated by Rosenstark [4], fills this requirement, and has the advantage that it can be carried out by hand calculation. They showed that for an *RC* network, the poles can be found from the polynomial

$$\Delta(s) = a_0 + a_1 s + a_2 s^2 + \cdots + a_n s^n \tag{5.30}$$

This can be obtained for a resistive network to which n capacitors are connected, as shown in Fig. 5.9, by evaluating a number of driving point

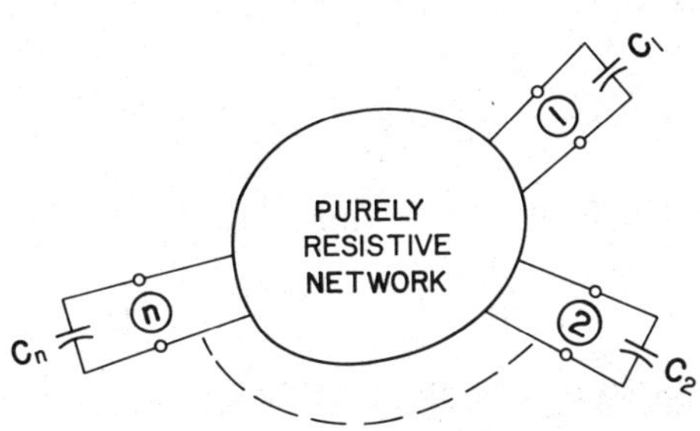

Figure 5.9 Resistive network with external capacitors.

resistances. The coefficients of (5.30) are given in terms of the time constants

$$a_0 = 1 \tag{5.31}$$

$$a_1 = \sum_{i=1}^{n} R_i^0 C_i = \sum_{i=1}^{n} \tau_i^0 \tag{5.32}$$

$$a_2 = \sum_{i=1}^{n-1} \sum_{j=i+1}^{n} R_i^0 C_i R_j^i C_j = \sum_{i=1}^{n-1} \sum_{j=i+1}^{n} \tau_i^0 \tau_j^i \tag{5.33}$$

$$a_3 = \sum_{i=1}^{n-2} \sum_{j=i+1}^{n-1} \sum_{k=j+1}^{n} R_i^0 C_i R_j^i C_j R_k^{ij} = \sum_{i=1}^{n-2} \sum_{j=i+1}^{n-1} \sum_{k=j+1}^{n} \tau_i^0 \tau_j^i \tau_k^{ij} \tag{5.34}$$

where the following definitions apply:

C_i = capacitance connected to port i (5.35)

R_i^0 = driving point resistance at port i when all other ports are open circuited (5.36)

R_j^i = driving point resistance at port j when port i is short circuited and all other ports are open circuited (5.37)

R_k^{ij} = driving point resistance at port k when ports i and j are short circuited and all other ports are open circuited (5.38)

$\tau_k^{ij} = R_k^{ij} C_k$ = time constant at port k when ports i and j are short circuited and all other ports are open circuited (5.39)

To expedite the calculation of driving point resistances in transistor circuits, we need some additional formulas for the resistances seen between terminals 1 and 3 and also terminals 1 and 2, which appear in Fig. 5.6.

Define

$$R_A = R_e + r_\pi (1 + g_m R_e) \tag{5.40}$$

We immediately observe that R_A will be infinite for FETs. We also define

$$g'_m = \frac{g_m}{1 + (g_m + 1/r_\pi) R_e} \approx \frac{g_m}{1 + g_m R_e} \tag{5.41}$$

After some manipulation it is found that

$$R_{13} = R_L + (R_i \| R_A)[1 + g'_m R_L] \tag{5.42}$$

If terminal 1 is shorted to terminal 2, then

$$(R_{13})_{1\text{-}2\ \text{shorted}} = R_L + (R_i \| R_e) \tag{5.43}$$

For calculation of driving point resistances at terminals 1–2

$$R_{12} = r_\pi \left\| \frac{R_i + R_e}{1 + g_m R_e} \right. \tag{5.44}$$

If terminal 1 is shorted to terminal 3, then

$$(R_{12})_{1\text{-}3\text{ shorted}} = (1/g_m) \| r_\pi \| [R_e + (R_i \| R_L)] \tag{5.45}$$

One additional point needs to be made before the calculations proceed. In a two-transistor problem, which contains four capacitors, only the two smaller valued poles need to be found, since the other two poles are much larger than the first two. This is substantiated by the pole locations for the two-transistor amplifier of Fig. 5.3, which are given in (5.6). The number of poles of interest corresponds to the number of transistors in the circuit. So, for a two-transistor problem, we need to find only the coefficients a_0, a_1, and a_2. In a three-transistor problem we would also have to find a_3. This consideration reduces very substantially the amount of work needed to find $\Delta(s)$.

5.6 *First Example—the Two-Transistor Amplifier*

The calculations were performed on the amplifier of Figs. 5.2 and 5.3. The equivalent circuit is redrawn in Fig. 5.10 with the capacitors replaced with the circled numbers, which designate the ports used for calculating the driving point resistances. Any convenient port numbering order can be used, and the one appearing in Fig. 5.10 was chosen arbitrarily.

We follow definition (5.39) and supporting Eqs. (5.40)–(5.45) to fill in Table 5.3.

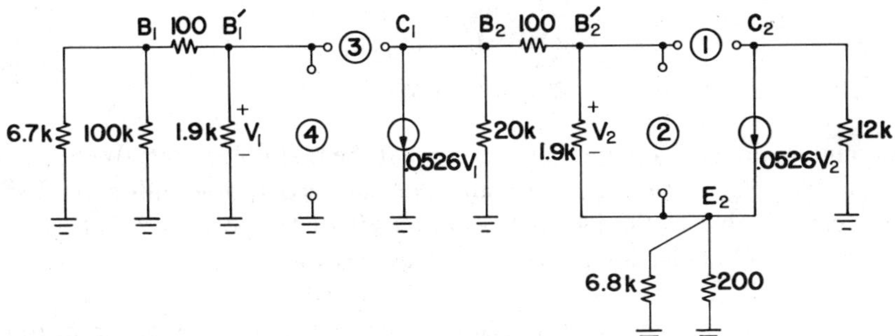

Figure 5.10 Amplifier of Fig. 5.3 ready for the application of the Cochrun–Grabel method.

Table 5.3 Computation schedule for the two-transistor amplifier of Fig. 5.10

(1)	(2)
$\tau_1^0 = 3036$	$\tau_2^1 = 1.5$ $\tau_3^1 = 126$ $\tau_4^1 = 117$
$\tau_2^0 = 74.1$	$\tau_3^2 = 120$ $\tau_4^2 = 117$
$\tau_3^0 = 4059$	$\tau_4^3 = 1.5$
$\tau_4^0 = 117$	

$a_0 = 1.$
a_1 = sum of all terms in column 1.
a_2 = sum of all row paired products.
Note: All time constants shown are in nanoseconds.

Finding the values for the table is not as formidable as it seems, because it is found through experience that the work is reduced by the fact that the driving point resistance at a port often remains unchanged, even though the conditions at other ports are varied.

Instead of using the rather intimidating (5.31)–(5.33) to evaluate the denominator polynomial coefficients, the rules given in Table 5.3 are followed. The second-order polynomial that results from the data of Table 5.3 is

$$\Delta(s) = 1 + 7.286 \times 10^3 s + 7.66 \times 10^5 s^2 \tag{5.46}$$

Keep in mind that the time constants given in Table 5.3 are in nanoseconds, hence the roots of (5.46) will be in gigaradians/second. The poles found from (5.46), when properly scaled, are

$$\begin{aligned} p_1' &= -0.139 \text{ Mrad/s} \\ p_2' &= -9.37 \text{ Mrad/s} \end{aligned} \tag{5.47}$$

A comparison of (5.47) and (5.6) shows that the first pole is exactly the same as that calculated using CORNAP, and that the second pole differs by 2.1%. This error is due to the fact that we used a reduced polynomial of second degree instead of the complete fourth-order polynomial.

The extra effort needed to obtain the fourth-order polynomial is not justified, as can be seen from an examination of Fig. 5.4. It is apparent from the fact that curve (a) hardly departs from curve (b), that the significant poles and zeros are enough to produce an accurate Bode plot at and beyond

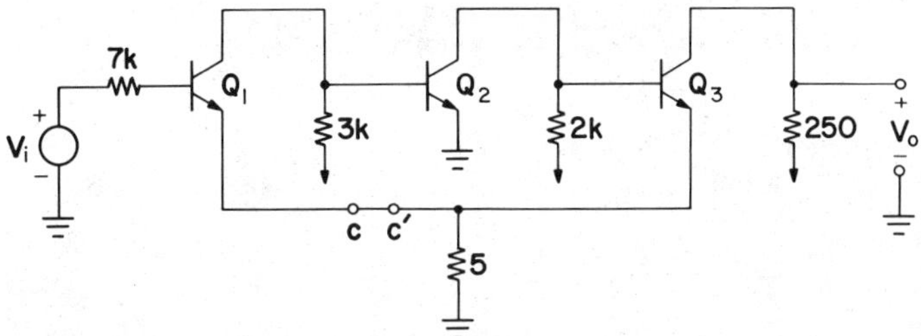

$r_x = 100 \quad r_\pi = 2k \quad C_\pi = 100pF \quad C_\mu = 5pF \quad g_m = .05\mho$

Figure 5.11 Three-transistor amplifier.

loop-gain crossover. If the zeros of the loop gain are ignored, then we see from curve (c) that there is a very significant error in the phase characteristic at loop-gain crossover. From curve (d) it becomes clear that if the Miller effect is used to calculate the poles (and the zeros are ignored), then the departure of this curve from curves (a) and (b) is so great that no useful information can be obtained at loop-gain crossover.

For the best results with the least effort, the principal poles should be calculated by the methods of Sec. 5.5, and the zeros should be found by the methods of Sec. 5.3. The need for compensation can now be investigated by any common method, including the Routh test (see Chap. 6), because this technique requires a knowledge of the transfer function critical points, which are now available.

5.7 *Second Example—the Three-Transistor Amplifier*

To offer further evidence of the principal pole and principal zero behavior in loop gain, we examine the loop-gain characteristic of the three-transistor amplifier of Fig. 5.11. The equivalent circuit prepared for the Cochrun–Grabel method is shown in Fig. 5.12.

The poles and zeros for the loop gain of the three-transistor amplifier were evaluated by using CORNAP, and it was found that the poles are located at

$$\begin{aligned} p_1 &= -0.927 \text{ Mrad/s} \\ p_2 &= -5.65 \text{ Mrad/s} \\ p_3 &= -36.9 \text{ Mrad/s} \\ p_4 &= -1313 \text{ Mrad/s} \\ p_5 &= -2510 \text{ Mrad/s} \\ p_6 &= -2942 \text{ Mrad/s} \end{aligned} \tag{5.48}$$

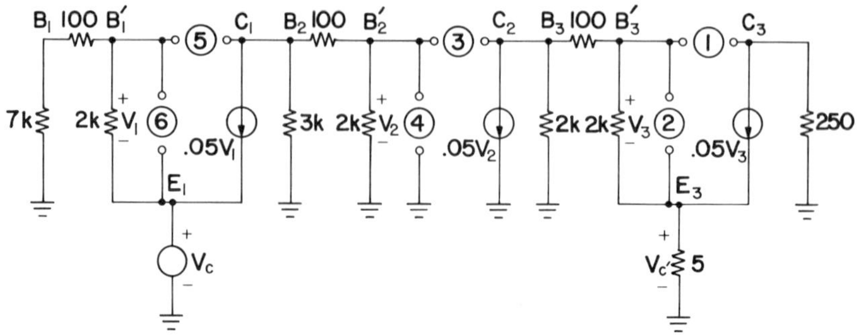

Figure 5.12 Three-transistor amplifier equivalent circuit.

The zeros are located at

$$\begin{aligned} z_1 &= -29.6 \text{ Mrad/s} \\ z_2 &= -475 \text{ Mrad/s} \\ z_3 &= -505 \text{ Mrad/s} \\ z_4 &= -800 \text{ Mrad/s} \\ z_5 &= +10^4 \text{ Mrad/s} \end{aligned} \tag{5.49}$$

The loop gain corresponding to these critical points is shown in Fig. 5.13, curve (a). Curve (b) is a plot of loop gain, which includes only the effects of p_1, p_2, p_3, and z_1. The phase characteristic shows only a departure of 6° at loop-gain crossover. This again supports the contention that the principal poles and zeros describe the loop-gain characteristic very adequately. Curve (c) shows that when loop-gain zeros are ignored, drastic compensation measures might be undertaken for an amplifier requiring only slight compensation. Curve (d) shows that the results obtained by using the Miller effect are of very little practical value for the study of stability in feedback amplifiers. All of the above points have already been made at the end of the previous section.

To calculate the poles independently, we use the definition (5.39) and supporting Eqs. (5.40)–(5.45) to fill in Table 5.4. The third-order polynomial that results from the data of Table 5.4 is

$$\Delta(s) = 1 + 1.285 \times 10^3 s + 2.27 \times 10^5 s^2 + 5.494 \times 10^6 s^3 \tag{5.50}$$

The time constants given in Table 5.4 are in nanoseconds, hence the roots of (5.50) will be in gigaradians/second. The poles found from (5.50), when properly scaled, are

$$\begin{aligned} p_1' &= -0.926 \text{ Mrad/s} \\ p_2' &= -5.66 \text{ Mrad/s} \\ p_3' &= -34.7 \text{ Mrad/s} \end{aligned} \tag{5.51}$$

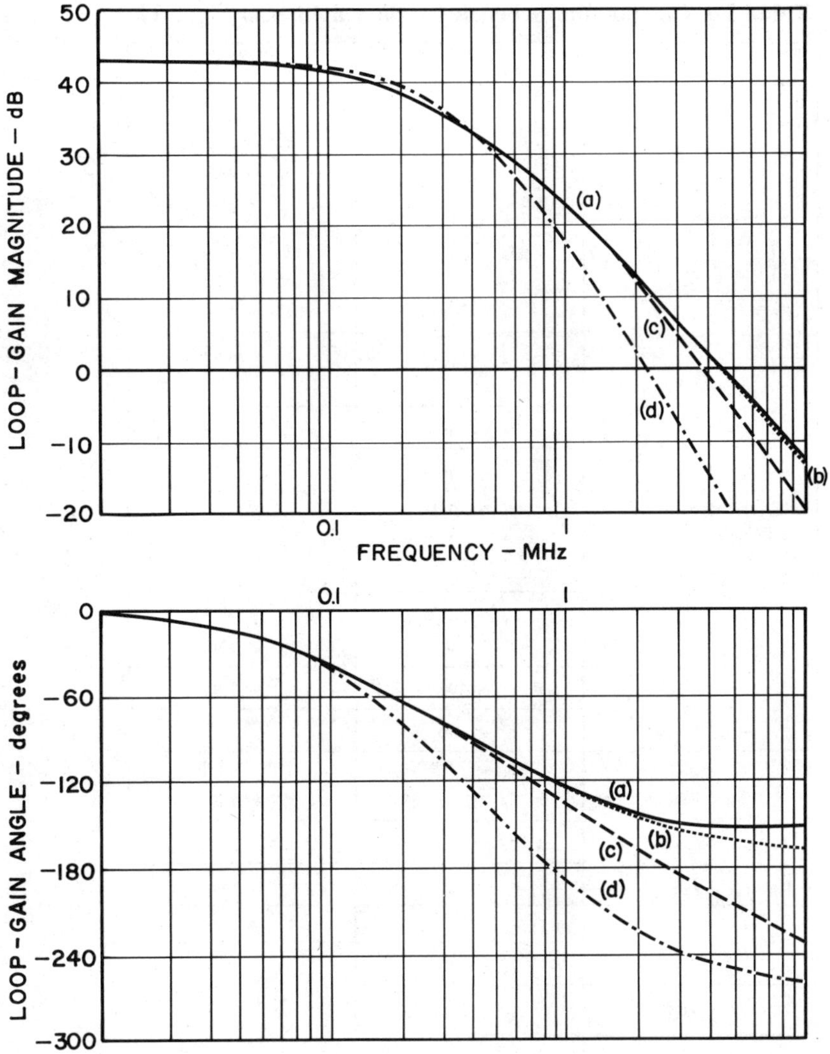

Figure 5.13 Loop-gain amplitude and phase characteristics. (*a*) Loop gain with all poles and zeros. (*b*) Loop gain with only significant poles and zeros. (*c*) Loop gain with significant poles only. (*d*) Loop gain with poles calculated using the Miller effect.

A comparison of (5.51) and (5.48) shows only a minor difference in the values of the poles calculated from (5.50). The small discrepancy is due to the fact that a reduced polynomial of third degree was used instead of the complete sixth-order polynomial.

The zeros were calculated by the methods of Sec. 5.3, and from the result it was determined that a principal zero is located at

$$z_1' = -29.6 \text{ Mrad/s} \tag{5.52}$$

Table 5.4 Computation schedule for the amplifier of Fig. 5.12

(1)	(2)	(3)
$\tau_1^0 = 64$	$\tau_2^1 = 2$	$\tau_3^{12} = 36.9$
		$\tau_4^{12} = 121.6$
		$\tau_5^{12} = 496$
		$\tau_6^{12} = 156$
	$\tau_3^1 = 41.8$	$\tau_4^{13} = 1.56$
		$\tau_5^{13} = 52.3$
		$\tau_6^{13} = 156$
	$\tau_4^1 = 121.6$	$\tau_5^{14} = 47.3$
		$\tau_6^{14} = 156$
	$\tau_5^1 = 496$	$\tau_6^{15} = 1.94$
	$\tau_6^1 = 156$	
$\tau_2^0 = 91.4$	$\tau_3^2 = 36.9$	$\tau_4^{23} = 1.53$
		$\tau_5^{23} = 52.1$
		$\tau_6^{23} = 156$
	$\tau_4^2 = 121.6$	$\tau_5^{24} = 47.3$
		$\tau_6^{24} = 156$
	$\tau_5^2 = 496$	$\tau_6^{25} = 1.94$
	$\tau_6^2 = 156$	
$\tau_3^0 = 355.5$	$\tau_4^3 = 1.93$	$\tau_5^{34} = 47.3$
		$\tau_6^{34} = 156$
	$\tau_5^3 = 53.2$	$\tau_6^{35} = 1.69$
	$\tau_6^3 = 156$	
$\tau_4^0 = 121.6$	$\tau_5^4 = 46$	$\tau_6^{45} = 1.64$
	$\tau_6^4 = 156$	
$\tau_5^0 = 496$	$\tau_6^5 = 1.94$	
$\tau_6^0 = 156$		

a_0 = 1.
a_1 = sum of all terms in column 1.
a_2 = sum of all row products of terms in columns 1 and 2.
a_3 = sum of all row products of terms in columns 1, 2 and 3.
Note: All time constants shown are in nanoseconds.

We see from Fig. 5.13 that the principal poles and zeros give a very close approximation to the exact frequency response of the loop gain of the feedback amplifier. If only principal poles are used, then this results in poor phase information at loop-gain crossover. The results obtained by using the Miller effect calculation are so poor as to be useless for the evaluation of performance parameters at loop-gain crossover.

5.8 *Conclusion*

It was shown that the two-transistor and three-transistor amplifiers have two and three significant loop-gain poles, respectively. Section 5.4 dealt with the Miller approximation and in Sec. 5.5 the very accurate Cochrun–Grabel method of hand-calculating the poles of any *RC* circuit was presented, and the method was demonstrated on a two-transistor feedback amplifier problem in Sec. 5.6. In Sec. 5.7 an example was given of the application of all these methods to a three-transistor feedback amplifier. The method of zero calculation was covered in Sec. 5.3 and applications were demonstrated in Secs. 5.3 and 5.7.

The examples showed that the results obtained compared very favorably with results obtained by using the powerful computer circuit analysis program CORNAP. It was shown, by use of examples, that the principal poles and zeros are needed to obtain adequate results for the loop-gain frequency response, and that ignoring the zeros produces a definitely inferior result. Finally, it was shown that calculations based on the Miller effect yield a totally inaccurate loop-gain frequency response. Blank schedules for the

Table 5.5 Blank computation schedule for two-transistor amplifiers

(1)	(2)
$\tau_1^0 =$	$\tau_2^1 =$ $\tau_3^1 =$ $\tau_4^1 =$
$\tau_2^0 =$	$\tau_3^2 =$ $\tau_4^2 =$
$\tau_3^0 =$	$\tau_4^3 =$
$\tau_4^0 =$	

$a_0 = 1.$
$a_1 =$ sum of all terms in column 1.
$a_2 =$ sum of all row paired products.

Table 5.6 Blank computation schedule for three-transistor amplifiers

(1)	(2)	(3)
$\tau_1^0 =$	$\tau_2^1 =$	$\tau_3^{12} =$ $\tau_4^{12} =$ $\tau_5^{12} =$ $\tau_6^{12} =$
	$\tau_3^1 =$	$\tau_4^{13} =$ $\tau_5^{13} =$ $\tau_6^{13} =$
	$\tau_4^1 =$	$\tau_5^{14} =$ $\tau_6^{14} =$
	$\tau_5^1 =$	$\tau_6^{15} =$
	$\tau_6^1 =$	
$\tau_2^0 =$	$\tau_3^2 =$	$\tau_4^{23} =$ $\tau_5^{23} =$ $\tau_6^{23} =$
	$\tau_4^2 =$	$\tau_5^{24} =$ $\tau_6^{24} =$
	$\tau_5^2 =$	$\tau_6^{25} =$
	$\tau_6^2 =$	
$\tau_3^0 =$	$\tau_4^3 =$	$\tau_5^{34} =$ $\tau_6^{34} =$
	$\tau_5^3 =$	$\tau_6^{35} =$
	$\tau_6^3 =$	
$\tau_4^0 =$	$\tau_5^4 =$	$\tau_6^{45} =$
	$\tau_6^4 =$	
$\tau_5^0 =$	$\tau_6^5 =$	
$\tau_6^0 =$		

$a_0 = 1$.
$a_1 =$ sum of all terms in column 1.
$a_2 =$ sum of all row products of terms in columns 1 and 2.
$a_3 =$ sum of all row products of terms in columns 1, 2 and 3.

calculation of the principal poles by application of the Cochrun–Grabel method to two-transistor and three-transistor amplifier problems appear in Tables 5.5 and 5.6. The supply of empty schedules can be increased endlessly by photocopying.

REFERENCES

1. Hayt, Jr. W. H., and Neudeck G. W.: "Electronic Circuit Analysis and Design," 2nd ed., Houghton Mifflin Co., Boston, 1984.
2. Miller, J. M.: Dependence of the Input Impedance of a Three-electrode Vacuum Tube upon the Load in the Plate Circuit, *Natl. Bur. Std. (U.S.) Res. Papers*, vol. 15, no. 351, pp. 367–385, 1919.
3. Cochrun, B. L., and Grabel, A.: A Method for the Determination of the Transfer Function of Electronic Circuits, *IEEE Transactions on Circuit Theory*, vol. 20, pp. 16–20, January, 1973.
4. Rosenstark, S.: Re-examination of Frequency Response Calculations for Feedback Amplifiers, *Int. J. of Electronics*, vol. 58, No. 2, pp. 271–282, February, 1985.

EXERCISES

5.1. For the one-transistor amplifier shown in Fig. P5.1:
(a) Find the gain at midband.
(b) Find the high-frequency performance using the Miller effect.

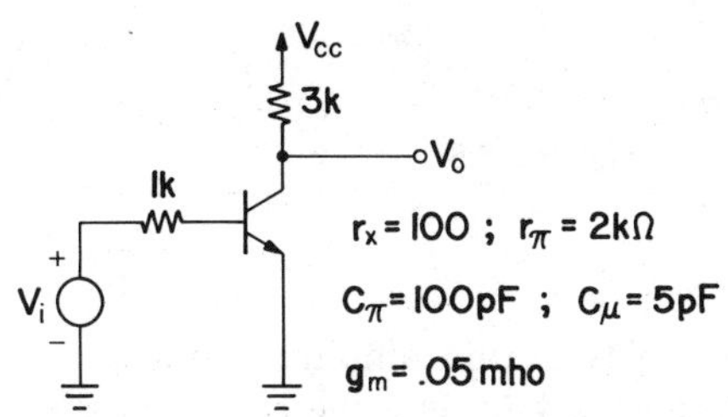

Figure P5.1

5.2. It is claimed that the cascode-connected amplifier shown in Fig. P5.2 has essentially the same gain as the one-transistor amplifier of Prob. 5.1, but it has a much better frequency response. Verify this by:
(a) Finding the midband gain.
(b) Calculating the high-frequency response using the Miller effect.

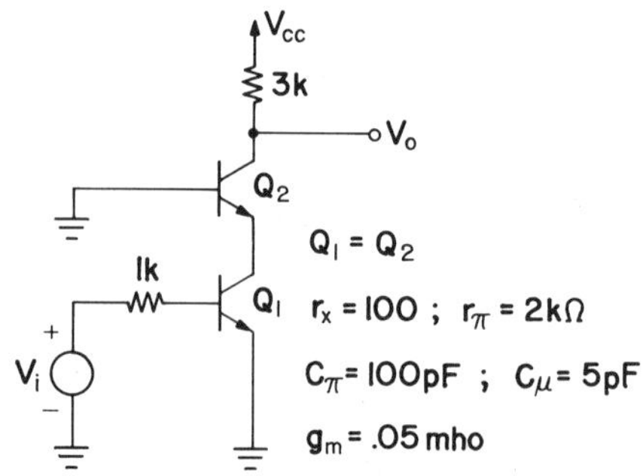

Figure P5.2

5.3. Find the zeros of the amplifier shown in Fig. P5.1.

5.4. Find the zeros of the amplifier shown in Fig. P5.2.

5.5. Find the poles of the source-free network shown in Fig. P5.5 by the following methods:
(a) Use the two-transistor schedule, but ignore all parts pertaining to ports 3 and 4.
(b) Write the admittance $Y(s)$ seen between any two convenient nodes. The zeros of $Y(s)$ should agree with part (a).
(c) Cut any convenient branch. Write the impedance $Z(s)$ seen between the resulting terminals of the cut. The zeros of $Z(s)$ should agree with parts (a) and (b).

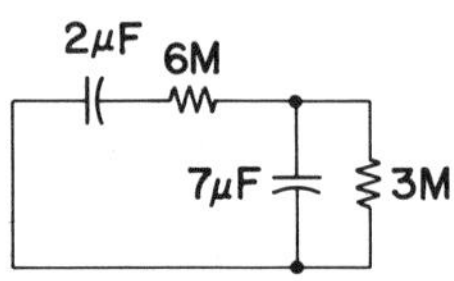

Figure P5.5

5.6. Find all the poles of the amplifier shown in Fig. P5.1 using the Cochrun–Grabel method. Use the two-transistor schedule, but ignore all parts pertaining to ports 3 and 4.

5.7. (a) Find the two principal poles of the amplifier shown in Fig. P5.2 using the Cochrun–Grabel method.
(b) How much of the schedule would have to be filled if it was only desired to find the first pole (approximately) of the amplifier.

5.8. Figure 5.3 is the equivalent circuit for finding the high-frequency loop gain of the two-stage amplifier of Fig. 5.2.
(a) Verify the values in Table 5.1.
(b) Verify the values in Table 5.2.
(c) Verify the values in Table 5.3.

5.9. For the amplifier equivalent circuit shown in Fig. 5.12, find the poles for each stage using the Miller approximation.

5.10. Starting with the amplifier equivalent circuit shown in Fig. 5.12, verify the values in Table 5.4.

6

Stability Analysis of Feedback Amplifiers

IN CHAP. 1 the substantial benefits that can be derived through the introduction of negative (degenerative) feedback into amplifiers were discussed, but nowhere was it mentioned that some difficulties may arise due to the use of feedback. Self-oscillation problems may arise in some cases; poor transient or frequency responses may arise in other cases. In this chapter a study will be made of the analytical methods needed to determine if an amplifier is stable, and Chap. 7 will take up the question of how to achieve a desired time or frequency response. A network will be considered stable if it has no tendency to oscillate. For stability it is necessary that the poles of the transfer function be located in the left-half s plane (LHP). The reader in need of a review of the Laplace transform concepts that are needed for an understanding of the last statement is referred to App. B.

6.1 *Stability of Feedback Circuits*

When dealing with the subject of electrical networks it becomes apparent that a requirement for stability is that all the poles of the transfer function must lie in the left half of the s plane. Since we are specifically concerned with feedback amplifiers, we have to see what part of the feedback expression it is that gives rise to the poles of the network, which will require close scrutiny to determine whether the feedback amplifier is stable or not.

Consider the asymptotic gain formula (2.18) for feedback amplifiers

$$G_f = G_\infty \frac{T}{1+T} + \frac{G_0}{1+T} \tag{6.1}$$

We must at first make allowance for the fact that all elements in the above equation could be frequency dependent, so that (6.1) will be written in the more general form

$$G_f(s) = \frac{G_\infty(s)T(s) + G_0(s)}{1 + T(s)} \tag{6.2}$$

An inspection of the various feedback amplifier configurations, which appear in Chap. 3, reveals that in most practical feedback amplifiers, G_∞ and G_0 are determined by passive elements, which are usually not of a frequency-dependent nature. At times a capacitor may appear in the feedback network, but that will be accounted for in the next chapter when phantom pole compensation is discussed. The closed loop gain $G_f(s)$ acquires its frequency dependence from $T(s)$ and the next equation reflects this fact.

$$G_f(s) = \frac{G_\infty T(s) + G_0}{1 + T(s)} \tag{6.3}$$

From (6.3) we see that the poles of $G_f(s)$ are due to the zeros of the denominator, which are the zeros of the return difference

$$F(s) = 1 + T(s) \tag{6.4}$$

For feedback amplifiers which can be modelled in terms of lumped, linear elements, $T(s)$ can be written in the rational form

$$T(s) = H\frac{N(s)}{D(s)} \tag{6.5}$$

Using this in (6.3), we see that the poles of $G_f(s)$ are due to the roots of the *characteristic equation*

$$C(s) = D(s) + HN(s) \tag{6.6}$$

The amplifier stability can be determined from the location of the roots of the above equation for situations in which $T(s)$ can be determined analytically. In that case the Routh–Hurwitz criterion can be used to check if any roots of the characteristic equation lie in the right-half s plane. There will be cases in which it is either too difficult to determine $T(s)$ accurately by analytical methods, or the designer has very little confidence in the model used to arrive at the analytical results. In such situations it is desirable to measure the frequency dependence of the return ratio T in the laboratory, and then the poles and zeros of (6.6) can only be estimated by curve fitting. This procedure introduces another element of doubt into the rational representation of $T(s)$. To circumvent all these problems, the Nyquist criterion is applied to the measured data directly, to determine the stability of the feedback amplifier. The two methods mentioned will be covered presently, but the Nyquist criterion is the more useful because it can be applied to analytical as well as laboratory results.

6.2 *The Routh–Hurwitz Criterion* [1–3]

The characteristic Eq. (6.6), is a polynomial in the complex variable s, which will be written in the form

$$C(s) = a_0 s^n + a_2 s^{n-1} + \cdots + a_{n-1} s + a_n = 0 \tag{6.7}$$

The coefficient a_0 is assumed positive. If it is not, then the test proceeds with a new characteristic equation which is the negative of the old one. To establish if any of the roots of $C(s)$ lie in the right-half s plane, we start by listing the polynomial coefficients in the top two rows of the array as shown below. If the polynomial is of even degree, then the polynomial coefficients of the even powers of s are listed in the first row, and the odd powers in the second row. Conversely, for a polynomial of odd degree:

$$\begin{array}{c|ccccc} s^n & a_0 & a_2 & a_4 & a_6 & \cdots \\ s^{n-1} & a_1 & a_3 & a_5 & a_7 & \cdots \\ s^{n-2} & b_1 & b_3 & b_5 & b_7 & \cdots \\ s^{n-3} & c_1 & c_3 & c_5 & c_7 & \cdots \\ \vdots & \vdots & & & \vdots & \cdots \end{array} \tag{6.8}$$

The coefficients below the second row are calculated using the following rules:

$$b_1 = \frac{a_1 a_2 - a_0 a_3}{a_1} \tag{6.9}$$

$$b_3 = \frac{a_1 a_4 - a_0 a_5}{a_1} \tag{6.10}$$

$$c_1 = \frac{b_1 a_3 - a_1 b_3}{b_1} \tag{6.11}$$

The rules for the other elements of the array can be inferred from the three rules given. The Routh–Hurwitz theorem states that for nondegenerate cases, the number of changes in sign of the first column of the array is equal to the number of roots of the polynomial $C(s)$. Degenerate cases will be covered later.

Example 6.1: Consider the characteristic equation

$$C(s) = s^4 + 10s^3 + 35s^2 + 50s + 24$$

The Routh–Hurwitz array is shown below

$$\begin{array}{c|ccc} s^4 & 1 & 35 & 24 \\ s^3 & 10 & 50 & \\ s^2 & 30 & 24 & \\ s^1 & 42 & & \\ s^0 & 24 & & \end{array}$$

There are no changes in sign in the terms in the first column of numbers of the array, so we conclude that $C(s)$ has no roots in the right-half plane. In fact

$$C(s) = (s+1)(s+2)(s+3)(s+4)$$ ■

Example 6.2: Consider the characteristic polynomial

$$C(s) = s^3 + 2s^2 + 2s + 40$$

The Routh–Hurwitz array for this polynomial is

$$\begin{array}{c|cc} s^3 & 1 & 2 \\ s^2 & 2 & 40 \\ s^1 & -18 & \\ s^0 & 40 & \end{array}$$

There are two changes of sign in the first column of the array, so this polynomial has two roots in the right-half s plane. ■

First Degenerate Case

A zero results in the first column of the nth row but no other element of the nth row is zero. If no remedial action is taken then the array cannot be continued. To circumvent this problem, we replace the zero in the nth row with a positive quantity ε, continue to fill the array, and then see what happens as ε approaches zero

Example 6.3: Consider the characteristic polynomial

$$C(s) = s^4 + s^3 + 2s^2 + 2s + 3$$

The Routh–Hurwitz array for this polynomial is

$$\begin{array}{c|ccc} s^4 & 1 & 2 & 3 \\ s^3 & 1 & 2 & \\ s^2 & \varepsilon & 3 & \\ s^1 & 2 - 3/\varepsilon & & \\ s^0 & 3 & & \end{array}$$

As ε goes to 0, there are two changes of sign in the first column of numbers, so this polynomial has two roots in the right-half s plane. It is obvious from the above that there will always be a change of sign in the first column of the array for this kind of situation, therefore if a zero appears in the first column the system is unstable. ■

Second Degenerate Case

An entire row becomes zero indicating an even power polynomial $C_e(s)$ is a factor of $C(s)$. The coefficients of the even polynomial correspond to the numbers in the row preceding the row of zeros. The system is distinctly unstable since polynomials of even power have roots in the s plane which are 180° apart. At the very least $C(s)$ will have roots on the $j\omega$ axis.

Example 6.4: Consider the characteristic polynomial

$$C(s) = s^3 + s^2 + 4s + A$$

The Routh–Hurwitz array for this polynomial is

$$\begin{array}{c|cc} s^3 & 1 & 4 \\ s^2 & 1 & A \\ s^1 & 4 - A & \\ s^0 & A & \end{array}$$

If $A = 4$ then the third row vanishes, and the even power polynomial coefficients are given by the elements of the second row of the array, so that

$$C_e(s) = s^2 + 4$$

which has roots at $s = \pm j2$. This system is unstable. ■

6.3 *The Encirclement Theorem*

The Nyquist stability criterion is based on the encirclement theorem. This topic will be the subject of discussion in this section. Consider a function $F(s)$ of the complex variable s, possessing a zero of order m at z, and a pole of order n at p, located within a contour C as shown in Fig. 6.1.

It will be assumed that there are no other critical points inside the contour C. Within the contour C, $F(s)$ can be written in the form

$$F(s) = \frac{(s - z)^m}{(s - p)^n} F_a(s) \tag{6.12}$$

where $F_a(s)$ possesses no poles or zeros within the contour C. Taking the

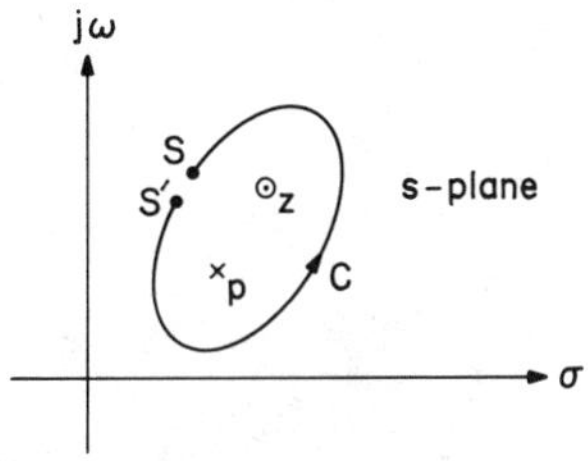

Figure 6.1 Contour C enclosing an mth-order zero and an nth-order pole.

natural logarithm of both sides of (6.12),

$$\ln[F(s)] = m\ln(s - z) - n\ln(s - p) + \ln[F_a(s)] \tag{6.13}$$

The above is differentiated with respect to s, to obtain

$$d\{\ln[F(s)]\} = \frac{m}{s - z}\,ds - \frac{n}{s - p}\,ds + \frac{F_a'(s)}{F_a(s)}\,ds \tag{6.14}$$

The last term on the right-hand side has no poles within the contour C, hence the right-hand side has the residue m at z and the residue $-n$ at p. Integrating (6.14) from S' to S along the contour C, results in

$$\ln[F(S)] - \ln[F(S')] = j2\pi(m - n) \tag{6.15}$$

Using the fact that $F(s)$ can be expressed in terms of magnitude and phase,

$$F(s) = |F(s)|e^{j\angle F(s)}$$

Then (6.15) can be rewritten into the form

$$\ln|F(S)| - \ln|F(S')| + j[\angle F(S) - \angle F(S')] = j2\pi(m - n) \tag{6.16}$$

As we close the gap in the contour C by allowing S to approach S', the first two terms in (6.16) cancel out, and we are left with

$$\angle F(S) - \angle F(S') = 2\pi(m - n) \tag{6.17}$$

The above result need not be restricted to one pole and one zero within the contour C. The derivation of the theorem is easily generalized. If C contains zeros of order $m_1, m_2, \ldots, m_r$ and poles of order $n_1, n_2, \ldots, n_q$, then we can say

$$\angle F(S) - \angle F(S') = 2\pi(M - N) \tag{6.18}$$

where

$$M = \sum_{i=1}^{r} m_i \tag{6.19}$$

denotes the total number of zeros within the contour C, and

$$N = \sum_{i=1}^{q} n_i \tag{6.20}$$

denotes the total number of poles within the contour. The left-hand side of (6.18) represents the change in angle of $F(s)$ as the variable s traverses the contour C in a counterclockwise direction. If both sides of (6.18) are divided by 2π, then the left-hand side represents the number of counterclockwise encirclements of the origin by $F(s)$, and this equals the right-hand side, which is the total number of zeros minus the total number of poles which $F(s)$ possesses inside the contour C. This statement is equally valid if the encirclements in the s plane and the $F(s)$ plane are both taken in a clockwise direction. The final statement of the encirclement theorem which leads to Nyquist's criterion is:

$$\begin{bmatrix} \text{The total number of encirclements of the} \\ \text{origin by } F(s) \text{ in a clockwise manner, as} \\ s \text{ traverses the contour } C \text{ in a clockwise} \\ \text{manner} \end{bmatrix} = \begin{bmatrix} \text{Total number of zeros} \\ \text{minus the total number} \\ \text{of poles within } C \end{bmatrix} \tag{6.21}$$

Example 6.5: Consider the function

$$F(s) = \frac{s+1}{s} \tag{6.22}$$

For the first contour in the s plane choose a circle of radius 0.8 around the origin as shown in Fig. 6.2a. This circle is described by

$$s = 0.8e^{j\theta}$$

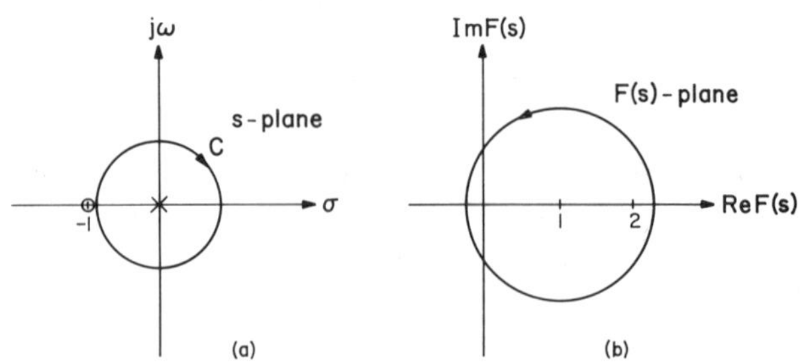

Figure 6.2 The mapping of the contour C by (6.22).

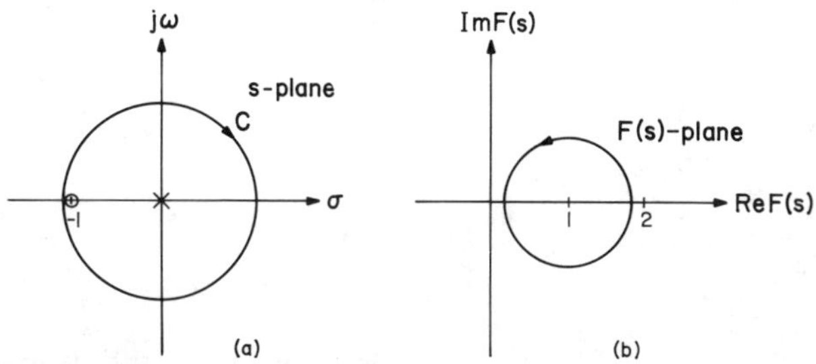

Figure 6.3 A different contour C mapped by (6.22).

When this is substituted into (6.22), the resultant expression is

$$F(s) = \frac{1 + 0.8e^{j\theta}}{0.8e^{j\theta}} = 1 + 1.25e^{-j\theta}$$

$F(s)$ traces out a circle of radius 1.25, centered on $+1$, in a direction opposite to that of contour C as shown in Fig. 6.2*b*. This encircles the origin -1 times, which is in agreement with (6.21), because C encircled only the pole of $F(s)$. When the circle for contour C is increased to a radius of 1.25, so that it encircles the pole and the zero, the resultant circle in the $F(s)$ plane does not encircle the origin, in agreement with (6.21). This is shown in Fig. 6.3. ■

6.4 *The Nyquist Stability Criterion*

Although it is desirable to determine the critical points of the return ratio accurately when the analysis of a feedback amplifier is carried out, the location of the critical points in the s plane cannot be verified directly by experiment. One could perform a loop-gain measurement, draw a Bode plot of the obtained data, and then attempt to fit the curves with asymptotes in order to determine the poles and zeros of the loop gain. This procedure is very error prone, and it is best to use a method for determining feedback amplifier stability, which uses the obtainable laboratory data directly. The Nyquist criterion, based on the theory of complex variables, is such a method.

The Nyquist criterion [4] addresses the question of whether (6.4) has zeros in the right-hand plane. At the outset it will be assumed that the system is open loop stable, so that $T(s)$, and therefore $F(s)$, have no poles in the right-half s-plane. Our only concern now is to discover if $F(s)$ has any zeros in the right-half s plane. To apply the encirclement theorem to the right-half s plane we choose a contour C as shown in Fig. 6.4.

As the radius of the circle R approaches infinity, the entire right-half plane will be enclosed. For the system to be stable, the image of C must not

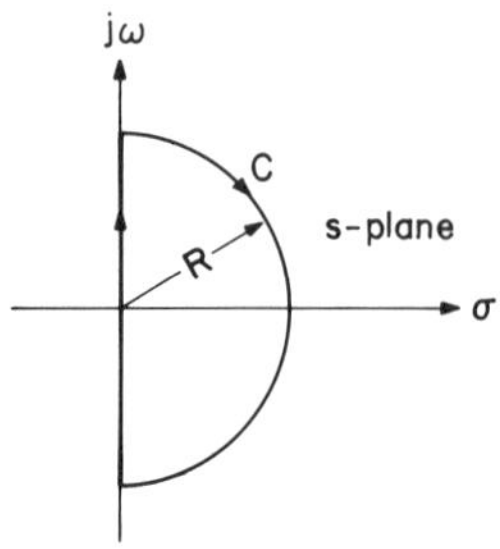

Figure 6.4 s plane contour for the Nyquist criterion.

encircle the origin in the $F(s)$ plane. This is equivalent to saying that the mapping of the contour C must not encircle the point $T(s) = -1$, since $F(s)$ and $T(s)$ are related by (6.4). It is more convenient in practice to work with the return ratio $T(s)$. In most practical feedback amplifiers, $T(s)$ is zero or a real constant at infinity, so that the large circle on the contour C usually contributes only one point to the Nyquist plot.

The steps needed to carry out the Nyquist criterion will now be summarized.

1. Ascertain that the amplifier is open loop stable. If it is not, then $T(s)$ has poles in the right-half s plane, and unless the number of poles can be determined, the Nyquist criterion becomes meaningless.
2. Using analytical or experimental methods, obtain data for $T(s)$ for $s = j\omega$. Plot this data in the $T(s)$ plane.
3. Determine if $T(s) = -1$ is encircled. In difficult cases, this is done by drawing a vector from $T(s) = -1$ to the contour plot of $T(j\omega)$. If the vector goes through a total of zero degrees as it follows the Nyquist plot from $\omega = -\infty$ to $\omega = \infty$, then the system is stable. Otherwise it is unstable.

Since $T(s)$ is the ratio of two polynomials in s with real coefficients, then $T(j\omega)$ is Hermitian, namely its amplitude and phase satisfy

$$
\begin{aligned}
|T(j\omega)| &= |T(-j\omega)| \\
\underline{/T(j\omega)} &= \underline{/-T(j\omega)}
\end{aligned}
\tag{6.23}
$$

Once the Nyquist plot is drawn for positive frequencies, the other half can be drawn in, or just imagined, since the plot is symmetrical around the horizontal axis. Before we proceed with examples, the concept of stability margins will be discussed.

6.5 *Stability Margins*

The Nyquist plot shown in Fig. 6.5 represents a stable feedback amplifier loop gain. If the point $T(s) = -1$ were encircled then an unstable amplifier

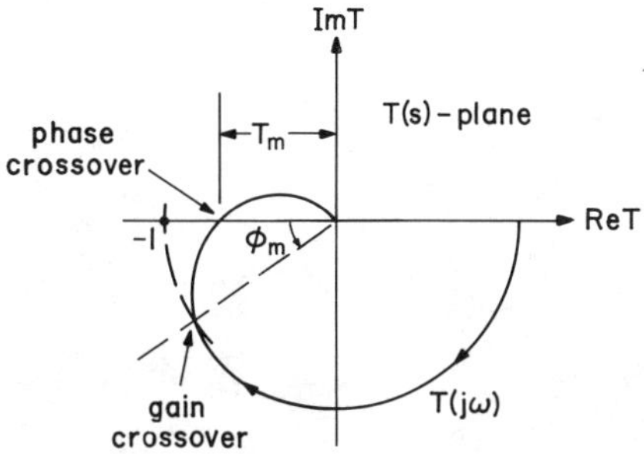

Figure 6.5 Nyquist plot for a hypothetical stable amplifier.

would result. The point $T(s) = -1$ represents the threshold of stability for the Nyquist locus. How close the Nyquist locus comes to that point is a reasonable indication of the margin of stability of the amplifier.

Phase-Crossover Frequency

This is the frequency at which the loop-gain phase goes through $-180°$. The reciprocal of T_m, the magnitude of the return ratio at this point, when expressed in dB, is the *gain margin* of the amplifier. The gain margin actually compares T_m to unity in an effort to establish how close the amplifier comes to oscillation when the return ratio phase goes through $-180°$. Gain margin is not defined for an amplifier whose phase never attains $-180°$.

Gain-Crossover Frequency

This is the frequency at which the loop-gain magnitude is unity (or 0 dB). The *phase margin* ϕ_m, is the value of phase attained at this point added to 180°. If the phase attained at gain crossover in our hypothetical case of Fig. 6.5 is $-120°$, then the phase margin is 60°. For most feedback amplifiers it is possible to relate step response performance to the phase margin (Chap. 7). For this reason the phase margin is a more important measure of the stability of feedback amplifiers than is the gain margin.

Example 6.6: The single pole feedback amplifier characteristic

$$T(j\omega) = \frac{T(0)}{1 + j\omega/\omega_0}$$

is not necessarily attributable to a single-stage amplifier. We have seen in the last chapter than even a single-stage amplifier transfer characteristic can contain two poles and one zero. Of the two poles, one will be

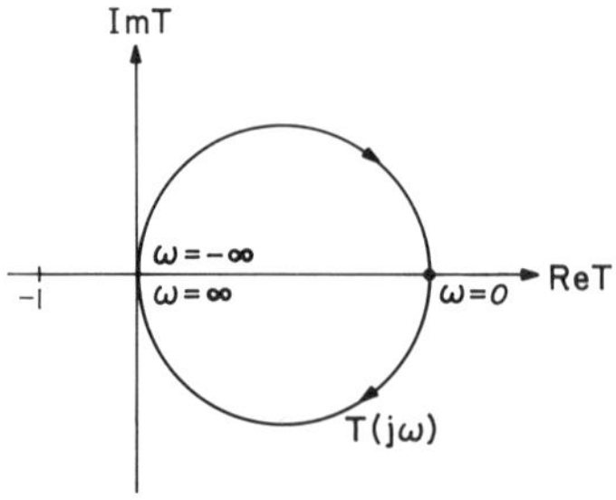

Figure 6.6 Nyquist plot for single pole amplifier.

significant. The zero may or may not be significant depending on the connection.

The Nyquist plot is a circle as shown in Fig. 6.6. The gain margin is not defined, and the phase margin is at least 90°. ■

Example 6.7: Determine if the amplifier whose Nyquist plot is shown in Fig. 6.7 is stable. Assume that the amplifier is open loop stable, namely $T(s)$ has no poles in the right-half s plane.

The simplest way to do this is to follow rule 3 of the Nyquist plot summary. The vector V shown in Fig. 6.7 follows the contour of $T(j\omega)$ as ω varies from $-\infty$ to $+\infty$. This vector swings through a total of 360°, so $1 + T(s)$ has one zero in the right-half s plane, and this amplifier is closed loop unstable. If it was not known that the amplifier is open loop stable, namely that $T(s)$ has no poles in the right-half s plane, then the Nyquist criterion would have been totally inconclusive. ■

Example 6.8: An amplifier with three poles in the loop gain has the return ratio given by

$$T(j\omega) = \frac{10}{(1 + j\omega/\omega_1)(1 + j\omega/20\omega_1)^2}$$

An attempt to plot the above on polar paper in order to produce a Nyquist locus is difficult because the return ratio magnitude changes too

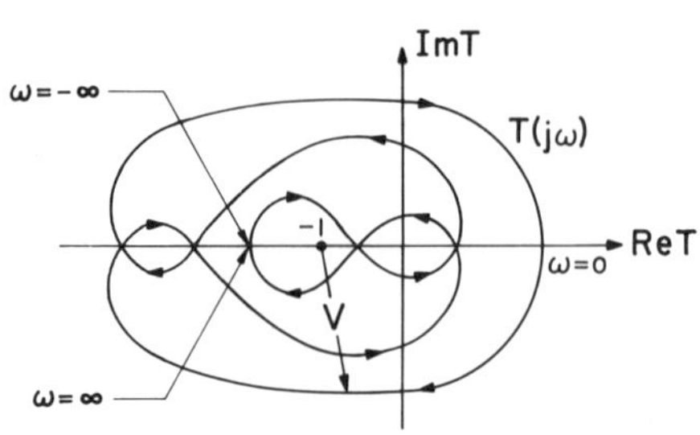

Figure 6.7 A somewhat convoluted Nyquist plot.

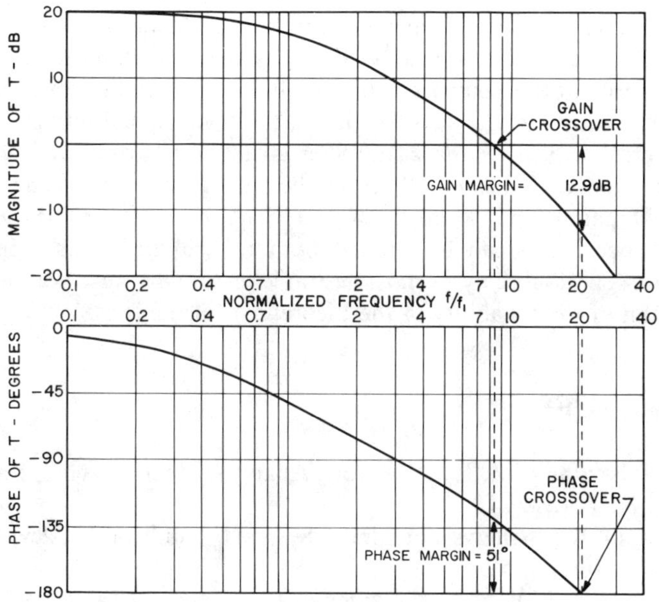

Figure 6.8 Bode plot for the three-pole amplifier.

much as frequency sweeps from 0 to ∞. The plot has very poor resolution in the region in the vicinity of $T(j\omega) = -1$, and it is therefore difficult to determine the stability margins accurately. The problem is circumvented when the data is presented on a Bode plot as shown in Fig. 6.8. It is very easy to see from this plot that for this amplifier, gain crossover occurs at 8.4 on the normalized frequency scale and the phase crossover occurs at 21. The gain margin is 12.9 dB, and the phase margin is 51°, and both were determined very accurately from the Bode plot. ■

It is apparent from the above example that the Nyquist criterion can be carried out on a Bode plot, and this is much easier to read than the polar plot needed for a Nyquist locus. Most feedback amplifiers usually do not have the convoluted Nyquist plot which is commonly found in control systems, so that the Bode plot facilitates the stability analysis of the amplifier, and gives better resolution in the region in which the stability margins have to be computed.

6.6 *Conclusion*

Two methods were presented for assessing if a feedback amplifier is stable. The Routh–Hurwitz test is very easy to carry out, but beyond determining if an amplifier is stable, it does not disclose just how stable the amplifier really is. This test relies on analytical results, and uncertainties in the model used to represent the elements of the amplifier, will affect the outcome of the Routh–Hurwitz test. On the other hand, the Nyquist stability criterion not only determines if a feedback amplifier is stable, but the stability margins which can be read from it disclose the quality of performance, as will become apparent in the next chapter. The Nyquist test is also more desirable because it can be applied both to analytical and experimental data. This is undoubtedly its most desirable feature, because it allows a final verification of the reliability of the theoretical calculations.

REFERENCES

1. Van Valkenburg, M. E.: "Network Analysis," Prentice-Hall, Englewood Cliffs, New Jersey, 1974.
2. Cheng, D. K.: "Analysis of Linear Systems," Addison-Wesley, Reading, Massachusetts, 1959.
3. Gantmacher, F. R.: "Applications of the Theory of Matrices," Interscience, New York, 1959.
4. Nyquist, H.: Regeneration Theory, *Bell System Technical Journal*, vol. 11, pp. 126–147, 1932.

EXERCISES

6.1. Just because an amplifier is open loop unstable does not necessarily mean that it is closed loop unstable. $T(s)$ appears in the numerator and denominator of (6.2).

(a) Prove that a pole of $T(s)$ is not a pole of $G_f(s)$.

(b) What is the expression for $\dfrac{T(s)}{1+T(s)}$ if $T(s) = \dfrac{2}{s-1}$.

(c) What is the expression for $\dfrac{T(s)}{1+T(s)}$ if $T(s) = \dfrac{s}{s-1}$.

6.2. Apply the Routh–Hurwitz test to the polynomials.

(a) $C(s) = s^4 + 26s^3 + 251s^2 + 1066s + 1680$.

(b) $C(s) = s^4 + 5s^3 + 2s + 10$.

(c) $C(s) = s^5 + 2s^4 + 2s^3 + 4s^2 + 11s + 10$.

(d) $C(s) = 3s^6 + s^5 + 19s^4 + 6s^3 + 81s^2 + 25s + 25$.

6.3. Consider the polynomial

$$C(s) = a_0s^4 + a_1s^3 + a_2s^2 + a_3s + a_4$$

Assume all coefficients are positive. Find a set of conditions, which the

```
100 REM   POLES AND ZEROS  : AMPLITUDE AND PHASE RESPONSE
110 DIM T(20),X(20),Y(20)
120 REM    The data is defined in terms of the form
130 REM    H(f) = H(1+jf/z1).../(1+jf/p1)... with complex
135 REM    conjugate critical points mentioned only once,
137 REM    their value given by their upper half s-plane
138 REM    location divided by 2*PI.  The pole at 2*PI*(
140 REM    -10+j100) would be stated below as -1,10,100.
145 REM
150 REM    Put into read tables the following:
160 REM
170 REM    Total number of zeros and poles.
180 DATA 3
190 REM
200 REM    Indicate poles by -1, then give location.
210 DATA -1,.159,0
220 DATA -1,.477,0
240 REM
250 REM    Indicate zeros by 1, then give location.
260 DATA 1,.318,0
280 REM
290 REM    Specify the constant H.
300 DATA 66.7
310 REM
320 REM    Specify the starting frequency and # of decades.
330 DATA .01,4
340 REM
350 READ M1
360 FOR M=1 TO M1
370   READ T(M),X(M),Y(M)
380 NEXT M
390 READ A1
400 A1=20*LOG(A1)/LOG(10)
410 READ F1,L
420 L=10*L
425 REM   Divides decades into 10 equal parts.  Can be changed.
430 S=10^.1
440 PRINT "NO."," FREQ","   AMP"," PHASE"
450 FOR K=0 TO L
460   A=A1
470   P=0
480   F=F1*S^K
490   FOR M=1 TO M1
500     IF Y(M)=0 THEN GOTO 560
510     C=F*Y(M)/(X(M)*X(M)+Y(M)*Y(M))
520     D=F*X(M)/(X(M)*X(M)+Y(M)*Y(M))
530     A=A+T(M)*10*LOG(((1+C)*(1+C)+D*D)*((1-C)*(1-C)+D*D))/LOG(10)
540     P=P+T(M)*(ATN(D/(1+C))+ATN(D/(1-C)))
550     GOTO 580
560     A=A+T(M)*10*LOG(1+(F/X(M))^2)/LOG(10)
570     P=P+T(M)*ATN(F/X(M))
580   NEXT M
590   PRINT K,F,A,P*180/3.14159
600 NEXT K
610 END
```

coefficients must meet if all the roots of the polynomial are to lie in the left half of the complex s plane.

6.4. Plot the Nyquist locus for the value $A = 1$, and from this determine the range of A for making the amplifier stable.

(a) $T(s) = A\dfrac{s-1}{s+1}$.

(b) $T(s) = A\dfrac{s+1}{s-1}$.

6.5. From the Bode plots of the given return ratios, determine the stability margins.

(a) $T(s) = \dfrac{100(s+2)}{(s+1)(s+3)}$.

(b) $T(s) = \dfrac{880}{(s+1)(s+2)(s+20)}$.

6.6. Determine the range of values of the midband gain $T(0)$ for which the feedback amplifier is stable. To do this obtain a Bode plot with $T(0) = 1$.

$$T(jf) = \frac{T(0)}{(1 + jf)^2(1 + jf/3)}$$

6.7. What is the value of $T(0)$ needed to get a phase margin of 64°?

$$T(jf) = \frac{T(0)(1 + jf/3)}{(1 + jf)(1 + jf/4)(1 + jf/5)}$$

The program appearing on the preceding page can be used to obtain values for the kinds of expressions found in Probs. 6.6 and 6.7. It should run without modification with most BASIC interpreters.

7

Feedback Amplifier Compensation

IN THE PREVIOUS CHAPTERS we addressed the questions of how to find the loop gain, the loop-gain frequency response, and how to determine the stability margins of a feedback amplifier from the frequency response data. This chapter will address the questions of how to relate the performance of the feedback amplifier in the time and frequency domain to its phase margin, and also how to alter the loop-gain characteristic of the amplifier to obtain a specific closed loop performance characteristic. To achieve our goal we shall try to relate the performance of feedback amplifiers to the performance of a standard second-order system.

7.1 *Second-Order System Response*

For this analysis we shall study a very simple second-order system possessing only two poles in the complex s plane. Its transfer function takes the form

$$H(s) = \frac{\omega_n^2}{s^2 + 2\zeta\omega_n s + \omega_n^2} \tag{7.1}$$

The parameter ζ is referred to as the damping factor, and ω_n is the natural frequency of the purely sinusoidal impulse response, which occurs when ζ is zero.

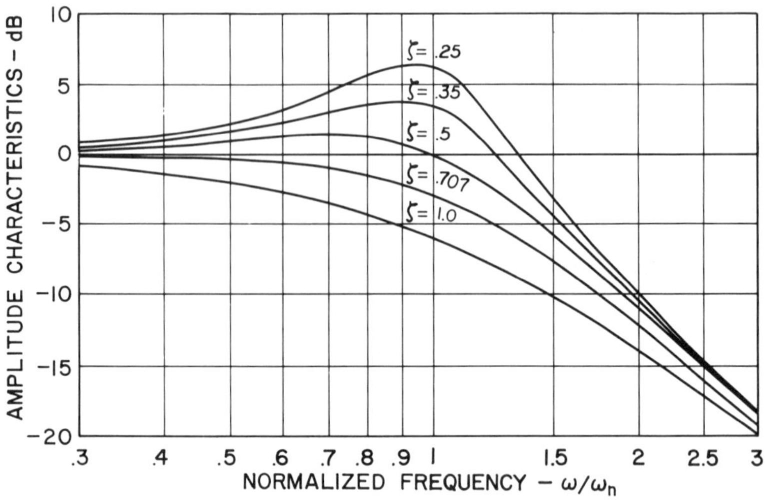

Figure 7.1 Second-order system amplitude response.

When s is replaced by $j\omega$, we obtain

$$H(j\omega) = \frac{1}{1 - (\omega/\omega_n)^2 + j2\zeta(\omega/\omega_n)} \tag{7.2}$$

The magnitude of the above is given by

$$|H(j\omega)|^2 = \frac{1}{\left[1 - (\omega/\omega_n)^2\right]^2 + 4\zeta^2(\omega/\omega_n)^2} \tag{7.3}$$

A plot of (7.3) for a few values of ζ appears in Fig. 7.1.

It is observed that for values of ζ which are $1/\sqrt{2}$ or greater, the amplitude characteristic has no overshoot. In fact, $\zeta = 1/\sqrt{2}$ divides the amplitude characteristics which have overshoot from those that do not. The characteristic for this value of ζ is referred to as the maximally flat response and it is in fact the Butterworth response of second order. The positions of the peaks in the amplitude response can be determined by setting to zero the derivative of the denominator of (7.3). The overshoot peak occurs at

$$(\omega/\omega_n)_{\text{peak}} = \sqrt{1 - 2\zeta^2} \tag{7.4}$$

When this is substituted into (7.3), the expression for the value of the amplitude response at the overshoot is found to have the form

$$|H(j\omega)|_{\text{peak}} = \frac{1}{2\zeta\sqrt{1 - \zeta^2}} \tag{7.5}$$

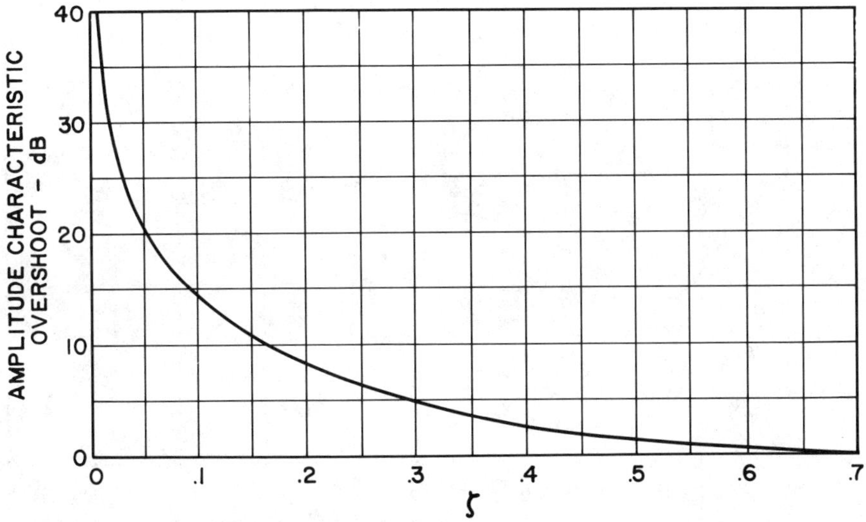

Figure 7.2 Amplitude response overshoot versus ζ.

A plot of the amplitude characteristic overshoot as a function of ζ is shown in Fig. 7.2.

The transfer function (7.1) has complex poles for $\zeta < 1$, and real poles for $\zeta \geq 1$. It will therefore have two different types of step response corresponding to those two cases. We are primarily interested in the underdamped case corresponding to $\zeta < 1$, and the critically damped case for $\zeta = 1$. The response to a unit step of a filter with the transfer function (7.1) for the underdamped case is

$$v_o(t) = 1 - e^{-\zeta\omega_n t}\left[\frac{\zeta}{\sqrt{1-\zeta^2}}\sin\sqrt{1-\zeta^2}\,\omega_n t + \cos\sqrt{1-\zeta^2}\,\omega_n t\right] \tag{7.6}$$

The same expression can be used to obtain results for the critically damped case, by assigning a value to ζ which is arbitrarily close to unity. Some typical step responses are shown in Fig. 7.3.

We see from Fig. 7.3 that the amount of step response overshoot is related to the value of the damping factor ζ. To find the position of the first peak we differentiate (7.6) with respect to $\omega_n t$ and set to zero, to find that the value where the first peak occurs is

$$(\omega_n t)_{\text{peak}} = \frac{\pi}{\sqrt{1-\zeta^2}} \tag{7.7}$$

When this result is substituted back into (7.6), the first peak is found to have

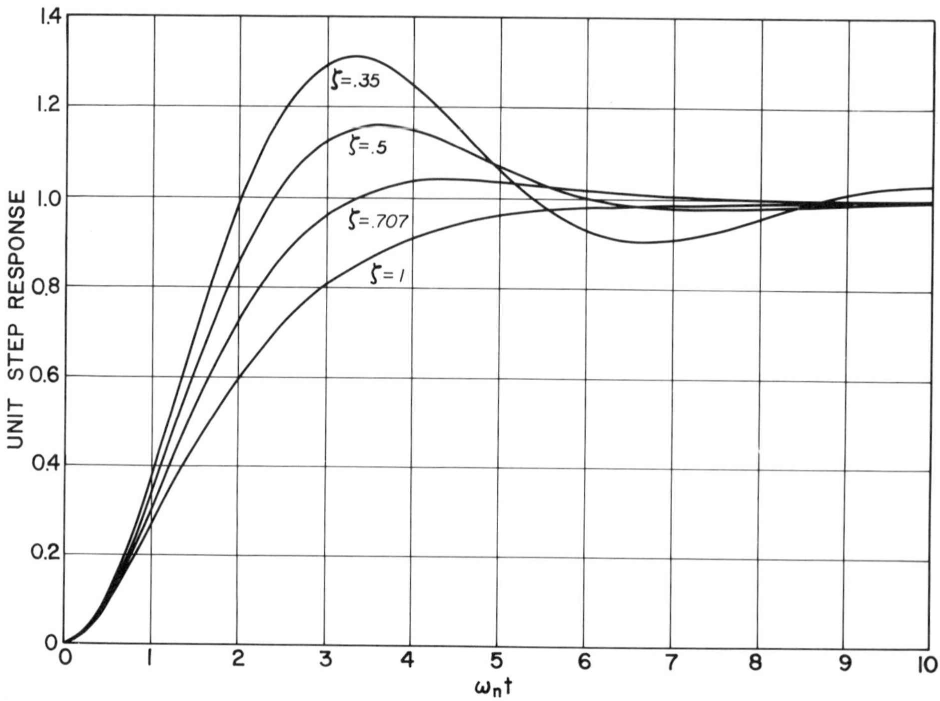

Figure 7.3 Response of the second-order system to a unit step input.

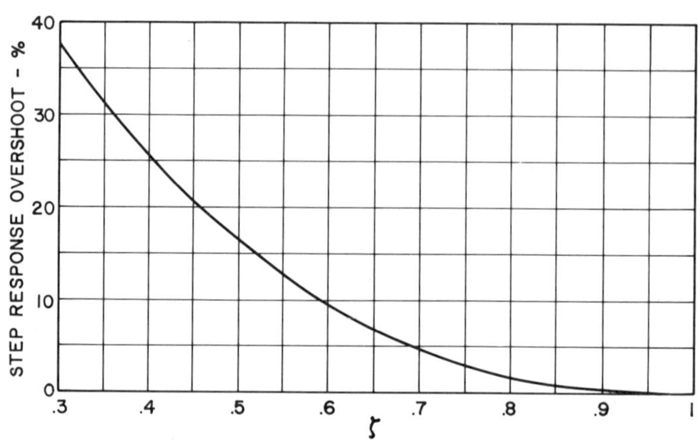

Figure 7.4 Step response overshoot versus ζ.

a value of

$$(v_o)_{\text{peak}} = 1 + e^{-(\zeta/\sqrt{1-\zeta^2})\pi} \tag{7.8}$$

The relative overshoot in the step response is therefore

$$\text{Step response overshoot} = e^{-(\zeta/\sqrt{1-\zeta^2})\pi} \tag{7.9}$$

A curve for (7.9) is shown in Fig. 7.4.

Now that we are familiar with the response of second-order systems, we can address the question of frequency and time behavior of second-order feedback systems.

7.2 *The Response of Second-Order Feedback Amplifiers*

Assume we are dealing with an amplifier whose loop gain possesses only two poles, as expressed in

$$T(s) = \frac{T_0 p_1 p_2}{(s + p_1)(s + p_2)} \qquad p_2 > p_1 \tag{7.10}$$

We define the pole separation factor

$$\frac{p_2}{p_1} = \alpha \equiv \text{pole separation factor} \tag{7.11}$$

Rewriting (7.10) using this definition, we get

$$T(s) = \frac{T_0 \alpha p_1^2}{(s + p_1)(s + \alpha p_1)} \qquad \alpha > 1 \tag{7.12}$$

The cases of interest are those in which

$$T_0 \gg 1 \tag{7.13}$$

It will be seen later that amplifiers for which T_0 is small do not usually have problems with stability, so that the above restriction does not diminish the scope of the stability analysis which follows.

What should be the relative spacing α of the two poles to obtain a desired frequency response or a desired transient response from the closed loop amplifier? To answer this question we shall examine the closed loop feedback amplifier response by comparing it to the standard second-order system response examined in Sec. 7.1. On the assumption that asymptotic gain is frequency independent and that the direct transmission term G_0 is so small that it will not have an effect on the closed loop response, we substitute (7.12) into (6.3), to obtain

$$G_f(s) = G_\infty(s)\frac{T(s)}{1 + T(s)} = \frac{G_\infty T_0 \alpha p_1^2}{s^2 + (1+\alpha)p_1 s + (1+T_0)\alpha p_1^2} \tag{7.14}$$

We write this in the standard second-order form, similar to (7.1)

$$G(s) = \frac{G_\infty T_0 \alpha p_1^2}{s^2 + 2\zeta\omega_n s + \omega_n^2} \tag{7.15}$$

When the denominators of (7.14) and (7.15) are compared, it is found that

$$\omega_n^2 = (1 + T_0)\alpha p_1^2 \tag{7.16}$$

and

$$\zeta^2 = \frac{(1+\alpha)^2}{4(1+T_0)\alpha} \tag{7.17}$$

If the midband loop gain T_0 is fixed, then the only alternative is to move the poles p_1 and p_2 in order to adjust the pole spacing ratio α, for the purpose of obtaining a response corresponding to some desired performance parameter ζ. This value of α is a solution of (7.17), which leads to the quadratic equation

$$\alpha^2 - \left[4\zeta^2(1+T_0) - 2\right]\alpha + 1 = 0 \tag{7.18}$$

Our interest will be in cases in which ζ is 0.5 or greater, so that we can see that an additional consequence of (7.13) is

$$4\zeta^2 T_0 \gg 1 \tag{7.19}$$

With the above inequality applied to (7.18), the resultant approximate solution for α is

$$\alpha \approx 4\zeta^2 T_0 \tag{7.20}$$

The above approximation has an error of 10% for $T_0 = 10$ and $\zeta = 0.5$. The approximation improves very rapidly for larger values of ζ and T_0. In order to determine the phase margin of a feedback amplifier designed to produce the performance corresponding to a given ζ, we need to calculate the loop-gain crossover frequency ω_c, which is the frequency at which the magnitude of the return ratio is unity. Substituting (7.20) into (7.12), setting $s = j\omega$, and equating the magnitude of $T(j\omega)$ to unity, results in the equation

$$(\omega/p_1)^4 + (1 + 16\zeta^4 T_0^2)(\omega/p_1)^2 - (T_0^2 - 1)16\zeta^4 T_0^2 = 0 \quad (7.21)$$

If (7.13) and (7.19) are satisfied, then the last equation reduces to

$$(\omega/p_1)^4 + 16\zeta^4 T_0^2(\omega/p_1)^2 - 16\zeta^4 T_0^4 \approx 0 \quad (7.22)$$

The above is solved by

$$(\omega_c/p_1)^2 \approx 8\zeta^4 T_0^2\left[-1 + \sqrt{1 + \frac{1}{4\zeta^4}}\right] \quad (7.23)$$

The phase margin is found by substituting $s = j\omega_c$ into (7.12) and evaluating the phase angle. The phase margin can then be calculated using

$$\phi_m \approx 180^\circ - \tan^{-1}(\omega_c/p_1) - \tan^{-1}(\omega_c/\alpha p_1) \quad (7.24)$$

If (7.19) is satisfied, then from (7.23) we see that

$$(\omega_c/p_1) \gg 1 \quad (7.25)$$

and (7.24) reduces to the slightly simpler form

$$\phi_m \approx 90^\circ - \tan^{-1}(\omega_c/\alpha p_1) \quad (7.26)$$

It would be useful to know where the 3-dB frequency ω_3 of the closed loop amplifier is located. We have to find the frequency at which the square of the magnitude of the denominator of (7.15) is equal to 2 times its value at $\omega = 0$, when s is replaced by $j\omega$. The equation that has to be solved is

$$\left[1 - (\omega/\omega_n)^2\right]^2 + 4\zeta^2(\omega/\omega_n)^2 = 2 \quad (7.27)$$

This leads to the quadratic equation

$$(\omega/\omega_n)^4 - 2(1 - 2\zeta^2)(\omega/\omega_n)^2 - 1 = 0 \quad (7.28)$$

which is solved by

$$(\omega_3/\omega_n)^2 = (1 - 2\zeta^2) + \sqrt{(1 - 2\zeta^2)^2 + 1} \quad (7.29)$$

To replace ω_n by p_1 in the above expression, we substitute (7.20) into (7.16), and simplify on the assumption that T_0 is much greater than unity, to obtain

$$(\omega_n/p_1)^2 \approx 4\zeta^2 T_0^2 \quad (7.30)$$

Table 7.1 Summary of two-pole feedback amplifier performance data

ζ	$\alpha = p_2/p_1$	ω_c/p_1	ω_3/ω_c	ϕ_m
0.5	T_0	$0.79T_0$	1.62	51.8°
0.6	$1.44T_0$	$0.86T_0$	1.60	59.2°
0.707*	$2T_0$	$0.91T_0$	1.55	65.5°
0.8	$2.56T_0$	$0.94T_0$	1.48	69.9°
0.9	$3.24T_0$	$0.96T_0$	1.40	73.5°
1.0	$4T_0$	$0.97T_0$	1.32	76.3°
1.1	$4.84T_0$	$0.98T_0$	1.26	78.6°
damping factor	ratio of poles	loop-gain crossover	closed loop 3-dB point	phase margin

*Second-order Butterworth response.

When this is substituted into (7.29), we obtain the final result

$$(\omega_3/p_1)^2 \approx 4\zeta^2 T_0^2\left[(1-2\zeta^2)+\sqrt{(1-2\zeta^2)^2+1}\right] \qquad (7.31)$$

Using (7.20), (7.23), (7.26), and (7.31), we are in a position to compile the data in Table 7.1. It is more useful to relate ω_3 to ω_c, so that we can have some idea of where the closed loop amplifier 3-dB point occurs in relation to the loop-gain crossover frequency. Accordingly, we divide the result of (7.31) by the result of (7.23) to obtain the results for Table 7.1.

7.3 *Compensation by Narrowbanding*

We shall restrict our discussion to compensation using capacitors. This is the more accepted method, since capacitors are smaller, lighter, less expensive, and more readily available than inductors, and they also do not produce any undesirable magnetic fields. For ease of reference in all future discussions, it will be assumed that active device Q_a is responsible for p_1, and active device Q_b is responsible for p_2, even though these devices might be labelled otherwise in the schematic diagram.

The simplest method of compensation is to lower the pole p_1, by shunting the base to emitter (or gate to source) terminals of Q_a with a capacitor. If the capacitance C_π' of this stage has been previously calculated, then the increase in capacitance C_π' required to lower the pole p_1 can be easily calculated. It is fortunate that this method involves the stage with the lower pole frequency p_1, since we know from Chap. 5 that this is the only pole which can be calculated fairly accurately using the Miller effect. The capacitance C_π' cannot be accurately determined for the second high-frequency pole p_2 by use of the Miller approximation, since this method assumes resistive loading for each stage, a condition that may not exist once the circuit starts exhibiting frequency-dependent behavior.

Example 7.1: For a two-pole feedback amplifier, the poles are $p_1 = 100$ kHz, $p_2 = 400$ kHz, and the capacitance C'_π for Q_a, the stage responsible for p_1, is 300 pF. The loop gain T_0 is 50. Use narrowbanding to obtain a second-order Butterworth response for the closed loop amplifier.

From Table 7.1 we see that the second-order characteristic of interest is that corresponding to $\zeta = 0.707$, which requires that the pole separation α must equal $2T_0$, so that the desired value of α is 100 for this case. The second pole cannot be increased above 400 kHz in value with capacitive compensation, so that the first pole must be moved to $p'_1 = 4$ kHz. The value of C'_π, which is 300 pF, has to be increased to a value of 7500 pF, so that the B–E terminals of Q_a will have to be shunted by a 7200-pF capacitor. This analysis ignores r_x, the ohmic resistance of the transistor base region, which is contained in the hybrid-pi transistor model, but it will be found that this assumption has a negligible effect on the outcome of the compensation. Final value adjustments should be performed in the laboratory anyway.

From Table 7.1 we can calculate the loop-gain crossover for the compensated amplifier

$$f_c = 0.91(50)4 \text{ kHz} = 182 \text{ kHz}$$

and the 3-dB frequency of the closed loop feedback amplifier

$$f_3 = 1.55(182 \text{ kHz}) = 282 \text{ kHz}$$

Suppose it is known that the voltage gain of Q_a in going from $V_{b'e}$ to V_{ce} is -30, then we could benefit from the Miller effect, and place a capacitor of 240 pF across the terminals B–C of Q_a. This is 30 times smaller than the 7200-pF compensating capacitor, which was determined above. Practical considerations of size, cost, and availability determine which capacitor should be used. ■

7.4 *Narrowbanding when Q_a Is Inaccessible*

It may happen that active device Q_a, at which the loop must be compensated, is inaccessible, as it might be in an operational amplifier in which only a few terminals are made available to the user. If a point is accessible in the amplifier where the resistance to ground is R_o, which remains frequency independent until the operating frequency exceeds p_1 by one or preferably two octaves, then compensation can be performed by shunting these terminals with the series RC combination shown in Fig. 7.5*a*.

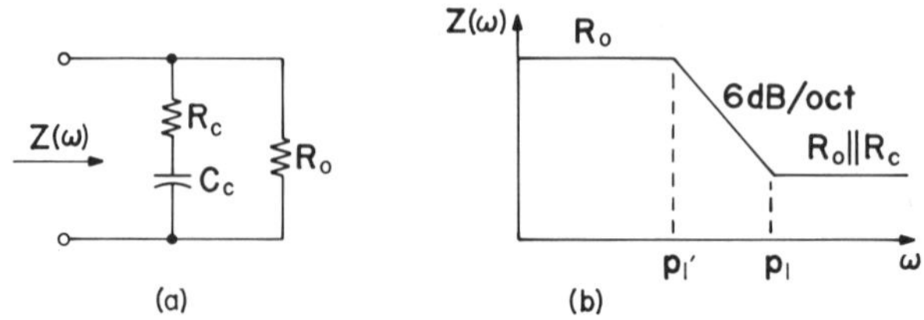

Figure 7.5 (*a*) Compensating network. (*b*) Its Bode asymptotes.

It can be seen from Fig. 7.5*b*, that this resultant impedance has a pole and a zero. If the zero is forced to occur at p_1 and the pole at p_1', then the zero will cancel the loop-gain pole at p_1, and the result will be that it will be replaced by the pole at p_1'.

The network zero is located at

$$p_1 = \frac{1}{R_c C_c} \tag{7.32}$$

and the pole at

$$p_1' = \frac{1}{(R_o + R_c)C_c} \tag{7.33}$$

The ratio of the two critical frequencies must equal γ, the pole relocation factor.

$$\frac{p_1}{p_1'} = \gamma \equiv \text{pole relocation factor} \tag{7.34}$$

Taking the ratio of (7.32) and (7.33), we find that R_c must satisfy

$$R_c = \frac{R_o}{\gamma - 1} \tag{7.35}$$

The value of C_c can now be determined from (7.32), and assuming that p_1 is specified in radians/second, then C_c is given by

$$C_c = \frac{1}{p_1 R_c} \tag{7.36}$$

Example 7.2: As in Example 7.1, $p_1 = 100$ kHz, and has to be relocated to 4 kHz, so that the pole relocation factor $\gamma = 25$. The active stage Q_a is inaccessible, but a point exists in the amplifier where the impedance to ground R_o is 2.4 kΩ, which remains frequency independent until the operating frequency is 400 kHz, which is two octaves above 100 kHz.

To compensate this amplifier, we can install a series RC branch across R_o. The value of the compensating resistor is found from (7.35)

$$R_c = \frac{R_o}{\gamma - 1} = \frac{2400}{25 - 1} = 100\ \Omega$$

The compensating capacitor C_c is calculated using (7.36), but it has to be kept in mind that p_1 is specified in Hertz and not in radians/second, hence

$$C_c = \frac{1}{p_1 R_c} = \frac{1}{2\pi(10^5)(100)} = 0.016\ \mu\text{F}$$

As was discussed at the end of Example 7.1, the compensating network can alternately be placed across the B–C terminals of Q_a, with the resistor needing a value of 3 kΩ, and the capacitor needing a value of 530 pF. ■

7.5 *Lag-Lead Compensation*

Narrowbanding is the least sophisticated of compensation methods, and its advantage lies in the fact that it is very easy to apply. With a little more trouble, a feedback amplifier can be compensated in a manner that does not reduce the bandwidth as much as narrowbanding does.

Assume as before that we are dealing with a two-pole feedback amplifier, and we wish to remove some area from the Bode plot as shown in Fig. 7.6*a*. This is an effort to relocate both poles to locations p_1' and p_2' to obtain the pole separation found in Table 7.1 for some desired performance specification. The cut in the Bode characteristic can be obtained by modifying the high-frequency response of Q_a, the stage responsible for the lower frequency pole p_1, to eliminate the existing pole at p_1, and in its place insert a pole at p_1', a zero at p_2, and a pole at p_2'. This modification is shown in Fig. 7.6*b*.

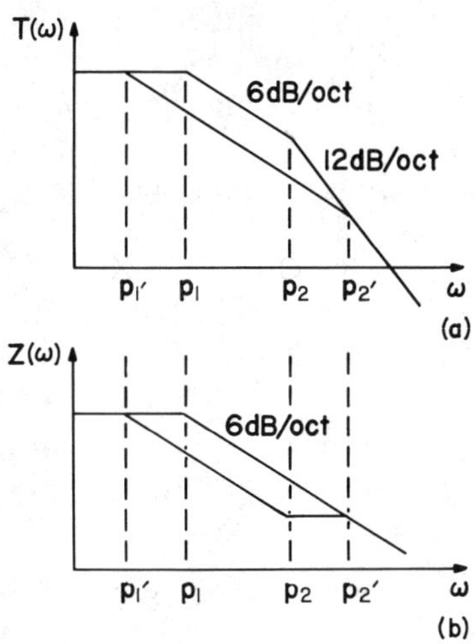

Figure 7.6 (*a*) Overall Bode characteristic. (*b*) Bode characteristic of stage Q_a.

Equal distances on the horizontal axis of a Bode plot represent equal ratios of frequencies. In connection with Fig. 7.6*b*, we define the pole relocation factor

$$\frac{p_1}{p_1'} = \frac{p_2'}{p_2} = \gamma \equiv \text{pole relocation factor} \tag{7.37}$$

The pole separation factor defined in (7.11) is the *desired* ratio of the poles p_1' and p_2' which has to be implemented in order to achieve a specific design (performance) objective.

$$\frac{p_2'}{p_1'} = \alpha \equiv \text{desired pole separation factor} \tag{7.38}$$

Using the last two definitions we can relate all critical frequencies to p_1, with the result

$$p_2' = \alpha p_1/\gamma \tag{7.39a}$$

$$p_2 = \alpha p_1/\gamma^2 \tag{7.39b}$$

$$p_1' = p_1/\gamma \tag{7.39c}$$

From (7.39b) we obtain

$$\gamma = \sqrt{\alpha p_1/p_2} \tag{7.40}$$

and this can be used to find γ for a given set of poles p_1, p_2, and a desired pole separation factor α.

The stage Q_a had at its input terminals B–E a parallel combination of a resistor r_π' and C_π', which for notational convenience will be referred to as R_1 and C_1, respectively. To compensate the feedback amplifier, this will be shunted by the series network consisting of R_c and C_c as shown in Fig. 7.7.

To see how compensation is achieved, we observe that prior to the introduction of the compensating network, the input impedance of this stage was

$$Z(s) = \frac{R_1 p_1}{s + p_1} \tag{7.41}$$

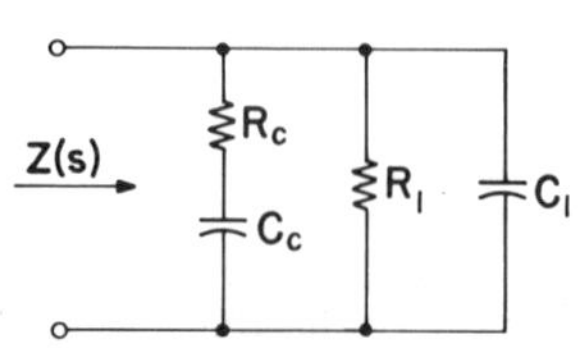

Figure 7.7 Compensation network to attain response shown in Fig. 7.6.

where

$$p_1 = \frac{1}{C_1 R_1} \tag{7.42}$$

We would like to modify the impedance function of (7.41), so that it will have the form

$$Z(s) = R_1 p_1 \frac{s + p_2}{(s + p_1')(s + p_2')} \tag{7.43}$$

With the compensating network in place, the loop-gain pole at p_2 will be cancelled, and new poles will be introduced at p_1' and p_2'. Using (7.39), the last equation can be rewritten in the form

$$Z(s) = R_1 p_1 \frac{s + \alpha p_1/\gamma^2}{(s + p_1/\gamma)(s + \alpha p_1/\gamma)} \tag{7.44}$$

To determine what values to assign to the components, we take the reciprocal of $Z(s)$ and divide by s in order to perform a partial fraction expansion

$$\frac{Y(s)}{s} = \frac{1}{R_1 p_1} \frac{(s + p_1/\gamma)(s + \alpha p_1/\gamma)}{s(s + \alpha p_1/\gamma^2)} = A + \frac{B}{s} + \frac{D}{s + \alpha p_1/\gamma^2} \tag{7.45}$$

The coefficients are readily evaluated as follows:

$$A = \left. \frac{Y(s)}{s} \right|_{s \to \infty} = \frac{1}{R_1 p_1} = C_1 \tag{7.46}$$

$$B = s \left. \frac{Y(s)}{s} \right|_{s=0} = \frac{1}{R_1} \tag{7.47}$$

$$D = (s + \alpha p_1/\gamma^2) \left. \frac{Y(s)}{s} \right|_{s = -\alpha p_1/\gamma^2} = \frac{(\alpha - \gamma)(\gamma - 1)}{\gamma^2 R_1} \tag{7.48}$$

With the above results we rewrite (7.45) in the form

$$Y(s) = C_1 s + \frac{1}{R_1} + \frac{(\alpha - \gamma)(\gamma - 1)}{\gamma^2 R_1} \frac{s}{s + \alpha p_1/\gamma^2} \tag{7.49}$$

On the right-hand side of (7.49), the first term is due to the capacitor C_1 and the second term is due to the resistor R_1, which were part of the original impedance in (7.41), so they need not be added to the compensating network. The last term on the right is the compensating admittance, so we write the reciprocal of it as the compensating impedance

$$Z_c(s) = \frac{\gamma^2 R_1}{(\alpha - \gamma)(\gamma - 1)} + \frac{\gamma^2 R_1}{(\alpha - \gamma)(\gamma - 1)} \frac{\alpha p_1/\gamma^2}{s} = R_c + \frac{1}{sC_c} \tag{7.50}$$

From the above equation it follows that

$$R_c = \frac{\gamma^2 R_1}{(\alpha - \gamma)(\gamma - 1)} \tag{7.51}$$

and by substituting (7.51) and (7.39b) into (7.50), we see that for p_2 in radians,

$$C_c = \frac{1}{p_2 R_c} \tag{7.52}$$

The above equations are used for calculating the values of the compensating elements needed in the series branch shown in Fig. 7.7.

Example 7.3: As in Example 7.1, $p_1 = 100$ kHz, $p_2 = 400$ kHz, and the desired pole separation factor α is 100. Since $C'_\pi = 300$ pF, then we can readily calculate that $r'_\pi = 5.3$ kΩ. In terms of the notation of this section

$$R_1 = r'_\pi = 5.3 \text{ k}\Omega$$

and

$$C_1 = C'_\pi = 300 \text{ pF}$$

The pole relocation factor is found using (7.40)

$$\gamma = \sqrt{\alpha p_1/p_2} = 5$$

We can now complete the design of the compensating network using (7.51)

$$R_c = \frac{\gamma^2 R_1}{(\alpha - \gamma)(\gamma - 1)} = \frac{25(5.3 \text{ k}\Omega)}{(100 - 5)(5 - 1)} = 349 \ \Omega$$

and we use (7.52) for C_c, but we keep in mind that p_2 is specified in Hertz, hence

$$C_c = \frac{1}{p_2 R_c} = \frac{1}{2\pi(4 \times 10^5)349} = 0.00114 \ \mu\text{F}$$

A consequence of this compensation technique is that pole p_1 is replaced by

$$p'_1 = p_1/\gamma = 20 \text{ kHz}$$

From Table 7.1 we find that loop-gain crossover occurs at

$$f_c = 0.91(50)20 \text{ kHz} = 910 \text{ kHz}$$

and the closed loop 3-dB bandwidth is

$$f_3 = 1.55(910 \text{ kHz}) = 1410 \text{ kHz} = 5(282 \text{ kHz})$$

The closed loop bandwidth of the amplifier in Example 7.1 was 282 kHz. We see that using this technique took a little more trouble, but we gained approximately two and one third octaves of closed loop bandwidth. The bandwidth has been improved by a factor γ. ■

7.6 *Compensation Using the Compensated Attenuator*

The methods of compensation discussed so far relied on the movement of the loop-gain poles. Pole p_1 is always moved down in frequency, although not as much in lag-lead compensation as in the case of narrowbanding. As a consequence, the closed loop bandwidth is reduced, since ω_3 is directly related to the value of the lowest pole p_1. The method to be presented in this section relies on changing the position of p_2, so that there is no reduction in closed loop bandwidth. The price, however, is a reduction in the midband loop-gain T_0.

If it is known at the outset that the amplifier will need to be compensated, then an attempt is made to obtain a midband loop-gain T_0 which is substantially greater than the one needed to attain the desired performance specification. An interstage circuit is then selected, as in Fig. 7.8*a*, and a resistive attenuator is introduced into the forward path to reduce the loop gain to the value needed to meet design objectives, as shown in Fig. 7.8*b*. The resistor R_2 and the capacitor C_2 represent the impedance between the transistor terminals B–E (or FET terminals G–S) of the device responsible for giving rise to p_2. The resistor R_0 is the collector (drain) load of the preceding stage. The Miller effect has to be considered when calculating C_2. The resistor r_x (in the bipolar-transistor model) is ignored in this analysis.

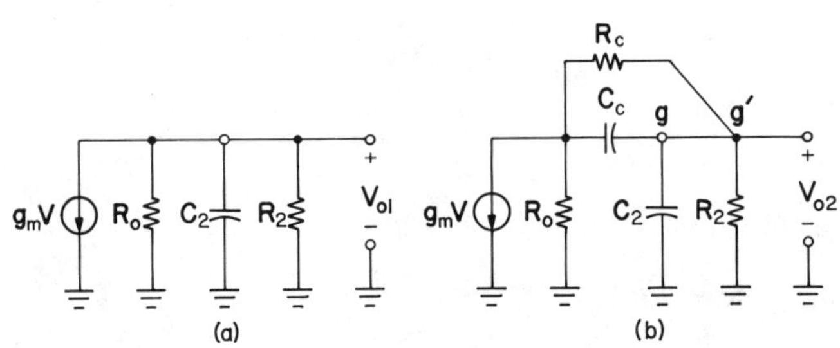

Figure 7.8 (*a*) Interstage before compensation. (*b*) Interstage after compensation.

The ratio of the voltage V_{01} to V_{02} at low frequencies represents the attenuation η introduced into the circuit and is given by

$$\eta = \frac{R_0 + R_c + R_2}{R_0 + R_2} \tag{7.53}$$

From this expression we find that

$$R_c = (R_0 + R_2)(\eta - 1) \tag{7.54}$$

The above can be used to calculate the value of R_c for a desired loop-gain attenuation.

The time constant of the interstage before the loop gain is compensated is obtained from Fig. 7.8*a*

$$\tau_2 = \frac{R_0 R_2}{R_0 + R_2} C_2 \tag{7.55}$$

and the break frequency, which this stage produces, in radians/second, is the reciprocal of the above

$$p_2 = 1/\tau_2 \tag{7.56}$$

If the capacitor C_c is not selected properly, then the interstage of Fig. 7.8*b* will produce two poles and one zero, which will make analysis very cumbersome. This problem can be avoided by selecting the capacitor to satisfy

$$C_c = \frac{R_2}{R_c} C_2 \tag{7.57}$$

This results in the bridge consisting of elements R_c, C_c, R_2, and C_2 being balanced, and the structure is referred to as a compensated attenuator. For this condition the operation of the circuit will not be affected by the removal of branch g–g', and we can finish the analysis by consulting Fig. 7.9.

The time constant of this circuit is

$$\tau_2' = \frac{R_0(R_c + R_2)}{R_0 + R_c + R_2} \frac{C_2 C_c}{C_2 + C_c} \tag{7.58}$$

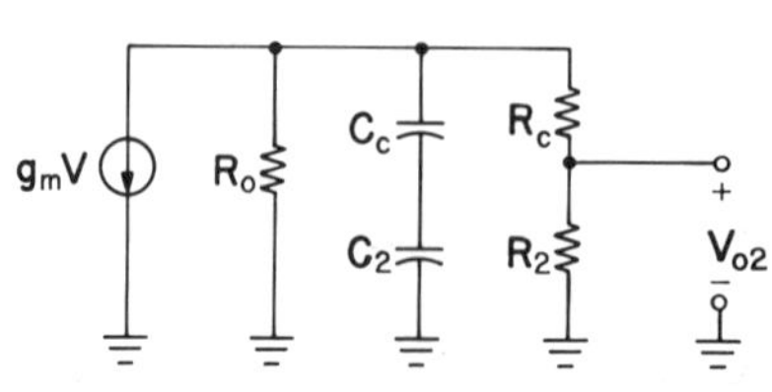

Figure 7.9 Equivalent circuit of Fig. 7.9*b*, when branch g–g' is removed.

When (7.57) and (7.54) are substituted into the above equation, and then a comparison is made with (7.55), it is found that

$$\tau_2' = \tau_2/\eta \tag{7.59}$$

and from the reciprocal of the above it is concluded that the new break frequency for this interstage is

$$p_2' = \eta p_2 \tag{7.60}$$

Example 7.4: It is desired to get a second-order Butterworth response for a feedback amplifier. At the outset, $p_1 = 1$ Mrad/s, $p_2 = 15$ Mrad/s, and $T_0 = 30$. T_0 exceeds specifications by at least a factor of 2. Compensate using the compensated attenuator method.

An examination of the amplifier reveals that the interstage, which gives rise to p_2, is the one shown in Fig. 7.10*a*. If the loop gain is reduced by a factor $\eta = 2$, to $T_0 = 15$, then the pole p_2 can be relocated to $p_2' = 30$ Mrad/s, which is η times its previous value. Now the pole separation factor α, which is the ratio of p_2' to p_1, will be 30. We know from Table 7.1 that this value of α is just right to get the desired Butterworth response.

The value of the compensating resistor R_c is found using (7.54)

$$R_c = (R_0 + R_2)(\eta - 1) = (2 + 1)(2 - 1) = 3\ \text{k}\Omega$$

and the value of the compensating capacitor is found using (7.57)

$$C_c = (R_2/R_c)C_2 = (1/3)100\ \text{pF} = 33.3\ \text{pF}$$

The compensated interstage is shown in Fig. 7.10*b*. At this point it is a good idea to recalculate the loop-gain frequency response. The compensation caused a change in loading of the preceding stage, and this could very well have caused a shift in its break frequency. ■

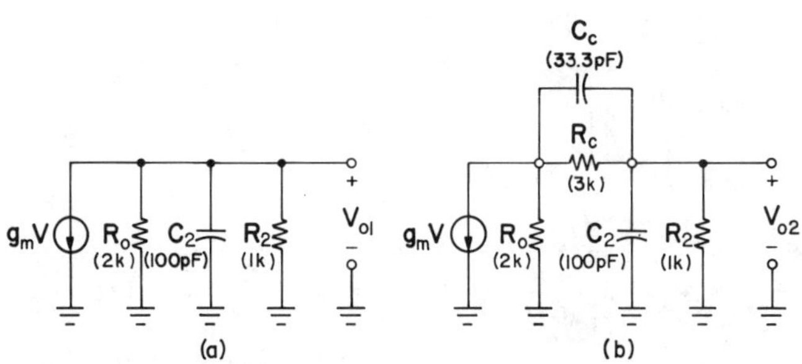

Figure 7.10 (*a*) Interstage before compensation. (*b*) Interstage after compensation.

7.7 *Phantom Zero Compensation*

The most desirable method of compensation is one which costs the designer nothing in bandwidth and which also does not affect the magnitude of the low-frequency loop gain T_0. The phantom zero compensation is a way of introducing a zero into the feedback loop, without having to first reduce the loop gain to make provision for this zero beforehand. It also does not disturb the location of the loop-gain pole p_1, so that it does not in any way narrowband the amplifier. If an amplifier has a large asymptotic gain G_∞, then we know from the formulas for asymptotic gain found in Chap. 3, that the resistors which are responsible for the value of asymptotic gain, also introduce substantial attenuation into the expression for return ratio. Bypassing this attenuation at high frequencies should introduce a zero into the expression for loop gain, and this should be helpful in correcting phase-margin deficiencies.

The amplifier shown in Fig. 7.11 has, in the absence of C_f, a return ratio $T_a(s)$ whose frequency behavior depends on $A(s)$.

$$T_a(s) = \frac{R_i'}{R_0 + R_f + R_i'} A(s) \tag{7.61}$$

where

$$R_i' = R_i \| R_s \tag{7.62}$$

The simplifying assumption will be made that

$$R_0 \ll R_f + R_i' \tag{7.63}$$

so that the return ratio can be written in the simpler form

$$T_a(s) = \frac{R_i'}{R_f + R_i'} A(s) \tag{7.64}$$

In the absence of C_f, the low-frequency asymptotic gain for the amplifier in Fig. 7.11 is given by

$$G_\infty(0) = -\frac{R_f}{R_s} \tag{7.65}$$

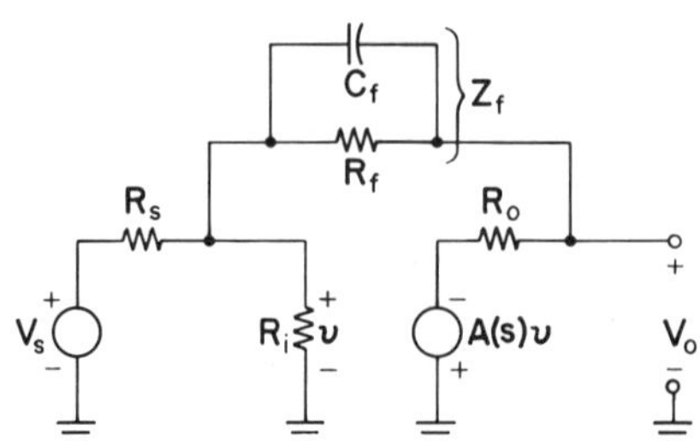

Figure 7.11 A model of a feedback amplifier.

If C_f is shunting R_f, then we replace R_f with

$$Z_f(s) = \frac{R_f}{R_f C_f s + 1} \tag{7.66}$$

in (7.64), to obtain

$$T(s) = \frac{R_i'}{Z_f + R_i'} A(s) = \frac{(R_f C_f s + 1)}{\left[\dfrac{R_f R_i' C_f}{R_f + R_i'} s + 1\right]} \frac{R_i'}{R_f + R_i'} A(s) \tag{7.67}$$

After consulting (7.64) we see that the above can be written in the form

$$T(s) = \frac{s/z_f + 1}{s/p_f + 1} T_a(s) \tag{7.68}$$

where

$$z_f = 1/R_f C_f \tag{7.69}$$

and

$$p_f = \frac{R_f + R_i'}{R_f R_i' C_f} = \left[1 + \frac{R_f}{R_s} \frac{R_s}{R_i'}\right] \frac{1}{R_f C_f}$$

Substitution into the above from (7.65) and (7.69) yields

$$p_f = \left[1 + |G_\infty(0)| \frac{R_s}{R_i'}\right] z_f \tag{7.70}$$

From (7.62) we know that the ratio $R_s/R_i' \geq 1$, so we conclude that

$$p_f \geq \left[1 + |G_\infty(0)|\right] z_f \tag{7.71}$$

Even if the magnitude of the low-frequency asymptotic gain $G_\infty(0)$ is small, then this method of compensation is still useful. In the case of a voltage follower, in which $|G_\infty(0)| = 1$, even if it is assumed very conservatively that the ratio of R_s to R_i' is unity, then the pole p_f will be $2z_f$, and this distribution of critical points is capable of giving a maximum phase correction of 19.5°. In most cases, a larger phase correction can be obtained.

The analysis which follows will be made much more tractable if it is assumed that the pole p_f is so far removed from all other critical points of the return ratio $T(s)$, that it can be disregarded, so that in place of (7.68), the expression that will be used for $T(s)$ is

$$T(s) = (s/z_f + 1) T_a(s) \tag{7.72}$$

When the compensating capacitor C_f is present, then in place of (7.65), we use for the asymptotic gain

$$G_\infty(s) = -\frac{R_f}{R_s}\frac{1}{C_f R_f s + 1}$$

which after substitution of (7.65) and (7.69), becomes

$$G_\infty(s) = G_\infty(0)\frac{1}{s/z_f + 1} \tag{7.73}$$

The closed loop response is obtained by substituting (7.72) and (7.73) into the asymptotic gain formula (7.14), with the result

$$G_f(s) = \frac{G_\infty(0)T_a(s)}{1 + (s/z_f + 1)T_a(s)} \tag{7.74}$$

We see that the return ratio zero, which would normally be a zero of $G_f(s)$, has been cancelled by the pole of $G_\infty(s)$, which is located at the same frequency. The vanishing of the zero is the reason that this is called the "phantom zero" method of compensation.

We again assume that we are dealing with a two pole feedback amplifier, so we substitute (7.10) into (7.74), with the result

$$G_f(s) = \frac{G_\infty(0)T_0 p_1 p_2}{s^2 + (p_1 + p_2 + T_0 p_1 p_2/z_f)s + (1 + T_0)p_1 p_2} \tag{7.75}$$

We could obtain any desired second-order response with the above equation, but to keep the analysis simple, an attempt will be made at obtaining a second-order Butterworth response, one corresponding to a damping factor $\zeta = 0.707$. For this response we require the terms in the denominator to satisfy

$$\sqrt{2}\,\omega_n = p_1 + p_2 + T_0 p_1 p_2/z_f \tag{7.76}$$

and

$$\omega_n^2 = (1 + T_0)p_1 p_2 \tag{7.77}$$

The last equation allows us to calculate ω_n, which for the case of the second-order Butterworth response is the 3-dB frequency of the closed loop amplifier. Once ω_n is determined, the location of the zero needed for proper compensation can be found by solving (7.76) for z_f

$$z_f = \frac{T_0 p_1 p_2}{\sqrt{2}\,\omega_n - (p_1 + p_2)} \tag{7.78}$$

The above has a solution as long as the denominator is positive. This is the case if

$$(p_1 + p_2)^2 < 2\omega_n^2$$

which on substitution of (7.77), can be reduced to

$$\frac{p_1}{p_2} + \frac{p_2}{p_1} < 2T_0 \qquad (7.79)$$

The above results in a quadratic equation in p_2/p_1. Solving of the quadratic equation can be avoided if it is assumed that

$$\frac{p_1}{p_2} \ll 1 \qquad (7.80)$$

in which case it follows directly from (7.79), that

$$\frac{p_2}{p_1} < 2T_0 \qquad (7.81)$$

As long as (7.81) is satisfied then (7.78) has solutions. This is not a limitation at all. If (7.81) is not satisfied, then the feedback amplifier has sufficient pole separation to have a response which corresponds to $\zeta \geq 0.707$, so that its performance is adequate without the need for further compensation. If compensation is needed then z_f should be calculated from (7.78). If a response which is more conservative than second-order Butterworth is desired, then the zero should be moved down in frequency. Moving the zero up in frequency results in a less conservative performance characteristic, with a step response which has a greater overshoot than the 4.3% of a second-order Butterworth response. The step response of the closed loop amplifier should be checked by computer calculation, or in the laboratory, to see if the desired performance characteristic has been obtained.

Example 7.5: As in Examples 7.1–7.3, it is assumed that an amplifier has poles at $p_1 = 100$ kHz and $p_2 = 400$ kHz. The low-frequency loop gain is 50. The low-frequency asymptotic gain is $G_\infty(0) = 10$ and $R_s/R_i' = 1$. The feedback resistor R_f has a value of 10 kΩ. Can phantom zero compensation be used to get a second-order Butterworth response?

Since all frequencies are stated in Hertz, we get from (7.77)

$$f_n = \sqrt{(1 + 50)(100)(400)} = 1428 \text{ kHz}$$

From (7.78) we find the location of the phantom zero

$$z_f = \frac{50(100)(400)}{\sqrt{2}\,(1428) - (100 + 400)} = 1316 \text{ kHz}$$

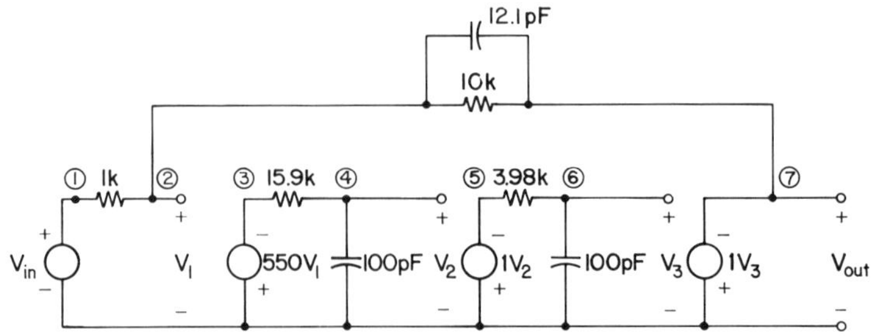

Figure 7.12 Configuration for simulation.

From (7.69) it is found that to obtain this zero, the feedback resistor of 10 kΩ has to be shunted by a capacitor of 12.1 pF. The pole which is associated with this method of compensation is found using (7.70)

$$p_f = [1 + 10(1)]1316 = 14.474 \text{ MHz}$$

and although it is a decade above the zero, it could have a significant influence on the phase margin, particularly if the loop-gain crossover frequency occurs above z_f. To examine this influence, we evaluate

$$T(jf) = \frac{50(1 + jf/1316)}{(1 + jf/100)(1 + jf/400)(1 + jf/14474)}$$

at a number of frequencies, to find that loop-gain crossover occurs at 1815 kHz and the phase margin is 62.5°. The pole at p_f reduces the phase margin by 7.16°. The phase margin is a trifle on the low side, so that we must check to see if we have an amplifier possessing a second-order Butterworth step response with an overshoot of 4.3%. To evaluate the transient response using the circuit analysis program SPICE on a VAX 11/780 computer, the amplifier was simulated using the configuration shown in Fig. 7.12.

This configuration meets all the specifications stated at the beginning of this example. The program needed to obtain the step response using the SPICE circuit analysis program is shown in Fig. 7.13.

The first computer run was made for the calculated capacitance C_f of 12.1 pF. Only a small portion of the output is presented in Fig. 7.13. Since the step response overshoot was 5.7%, a few additional trials were made with different values of C_f. A value of C_f of 12.5 pF (see Fig. 7.13*b*) produced the desired step response with an overshoot of 4.3%. The results are given in tabular form because the two columns of data are so close in value, that the difference between the two would have been almost imperceptible in a graph.

Step Response - 1st try

```
VIN 1 0 DC -0.102
RS 1 2 1K
RF 2 7 10K
CF 2 7 12.1PF
E1 0 3 2 0 550
R1 3 4 15.9K
C1 4 0 100PF
E2 0 5 4 0 1
R2 5 6 3.98K
C2 6 0 100PF
E3 0 7 6 0 1
.WIDTH IN=72 OUT=80
.OPTIONS LIST NOPAGE
.TF V(7) VIN
.TRAN 2E-8 40E-7 UIC
.PRINT TRAN V(7)
.END
```

TIME	V(7)	
6.000E-08	8.906E-02	
8.000E-08	1.546E-01	
1.000E-07	2.348E-01	
1.200E-07	3.150E-01	
1.400E-07	3.980E-01	
1.600E-07	4.812E-01	
1.800E-07	5.644E-01	
2.000E-07	6.399E-01	
2.200E-07	7.036E-01	
2.400E-07	7.673E-01	
2.600E-07	8.310E-01	
2.800E-07	8.833E-01	
3.000E-07	9.181E-01	
3.200E-07	9.530E-01	
3.400E-07	9.878E-01	
3.600E-07	1.014E+00	
3.800E-07	1.026E+00	
4.000E-07	1.038E+00	
4.200E-07	1.050E+00	
4.400E-07	1.057E+00	
4.600E-07	1.057E+00	<--peak
4.800E-07	1.056E+00	
5.000E-07	1.056E+00	
5.200E-07	1.053E+00	
5.400E-07	1.048E+00	
5.600E-07	1.043E+00	
5.800E-07	1.038E+00	
6.000E-07	1.033E+00	
6.200E-07	1.029E+00	

(a)

Step Response - 2nd try

```
VIN 1 0 DC -0.102
RS 1 2 1K
RF 2 7 10K
CF 2 7 12.5PF
E1 0 3 2 0 550
R1 3 4 15.9K
C1 4 0 100PF
E2 0 5 4 0 1
R2 5 6 3.98K
C2 6 0 100PF
E3 0 7 6 0 1
.WIDTH IN=72 OUT=80
.OPTIONS LIST NOPAGE
.TF V(7) VIN
.TRAN 2E 20E-7 UIC
.PRINT TRAN V(7)
.END
```

	TIME	V(7)
	6.000E-08	8.728E-02
	8.000E-08	1.550E-01
	1.000E-07	2.302E-01
	1.200E-07	3.132E-01
	1.400E-07	3.984E-01
	1.600E-07	4.813E-01
	1.800E-07	5.637E-01
	2.000E-07	6.375E-01
	2.200E-07	7.090E-01
	2.400E-07	7.692E-01
	2.600E-07	8.264E-01
	2.800E-07	8.719E-01
	3.000E-07	9.143E-01
	3.200E-07	9.462E-01
	3.400E-07	9.752E-01
	3.600E-07	9.956E-01
	3.800E-07	1.014E+00
	4.000E-07	1.025E+00
	4.200E-07	1.034E+00
	4.400E-07	1.039E+00
	4.600E-07	1.043E+00
peak-->	4.800E-07	1.043E+00
	5.000E-07	1.043E+00
	5.200E-07	1.041E+00
	5.400E-07	1.038E+00
	5.600E-07	1.035E+00
	5.800E-07	1.031E+00
	6.000E-07	1.027E+00
	6.200E-07	1.024E+00

(b)

Figure 7.13 SPICE data and output for two values of C_f.

The capacitance C_f can be increased further and the step response overshoot can be completely eliminated, but it is clear from the tabular data in Fig. 7.13, that a decrease in step response overshoot is accompanied by an increase in step response rise time. This indicates that there is also an attendant decrease in closed loop bandwidth. If this remained a purely second-order system after compensation, then the frequency f_n of 1428 kHz calculated above would be the closed loop 3-dB frequency. Since this is no longer a purely second-order system, it becomes necessary to evaluate the closed loop frequency response, by again applying SPICE to the configuration shown in Fig. 7.12. It was found that the bandwidth is 1.58 MHz for $C_f = 12.1$ pF and 1.5 MHz for $C_f = 12.5$ pF. Increasing C_f clearly reduces the amplifier bandwidth.

Phantom zero compensation is a very valuable method. The closed loop bandwidth of 1.5 MHz obtained here is greater than the 1.41 MHz for lag-lead compensation in Example 7.3. Since p_1 has not been changed by the process of compensation, the feedback bandwidth is a full 100 kHz in comparison to 20 kHz found in Example 7.3 and 4 kHz found in Example 7.1. ■

We see that phantom zero compensation should be the first choice, followed by lag-lead compensation, and narrowbanding should be used as a last resort. Naturally, if it is not necessary to squeeze the last ounce of performance out of the feedback amplifier, then narrowbanding is the easiest method to apply, and is also very simple to try in the laboratory. The stage thought to be responsible for pole p_1 gets a small capacitor shunted across its input terminals. If the amplifier performance degenerates, then this is not the stage responsible for p_1. If it improves, then the value of the capacitance is adjusted to give the desired performance.

It is possible to apply phantom zero compensation to any amplifier which has attenuation in its feedback path. If the impedances Z_{e1} and Z_{e2} are zero for the feedback amplifier of Fig. 3.4, then phantom zero compensation can still be obtained by putting a small inductor in series with the resistor, which would be found in place of Z_e. This is similar for the feedback amplifier of Fig. 3.8.

7.8 *Compensation of Higher-Order Amplifiers*

If feedback amplifiers which have more critical points than the two poles treated thus far are encountered, then a computer becomes a very useful tool for finding the best method of compensation. The analysis begins by finding the loop-gain frequency response by computer calculation or by experimental method. The compensation technique is independent of whether the amplifier is stable or unstable.

The loop-gain crossover frequency f_c is found first and the loop-gain phase ϕ_c, which is attained at that frequency, is then calculated. The relative spacing of p_f with respect to z_f is found from (7.70). The phantom zero method of compensation augments the loop gain with the transfer function

$$H(jf) = \frac{1 + jf/z_f}{1 + jf/p_f} \qquad p_f > z_f \tag{7.82}$$

which attains its maximum positive phase at the frequency

$$f_{\max \phi} = \sqrt{z_f p_f} \tag{7.83}$$

The value of this phase is

$$\phi_{\max} = \tan^{-1}\sqrt{\frac{p_f}{z_f}} - \tan^{-1}\sqrt{\frac{z_f}{p_f}} = 2\tan^{-1}\sqrt{\frac{p_f}{z_f}} - 90° \tag{7.84}$$

This is the maximum phase correction that could be obtained if $H(f)$ did not also affect the magnitude characteristic of the loop gain. It is therefore best to assume that $\phi_{\max}$ is 10° smaller than calculated. If

$$\phi_c + (\phi_{\max} - 10°) \geq -115° \tag{7.85}$$

then a 65°-phase margin can usually be achieved using only phantom zero compensation. If (7.85) is not satisfied, then lag-lead compensation should be used first to bring ϕ_c to a value which satisfies (7.85).

For a first try at phantom zero compensation, all but the two lowest-order poles of the loop gain $T(f)$ should be ignored. The value of z_f should be calculated using (7.77) and (7.78) and p_f should be obtained from (7.70). These should be included in the expression for $T(f)$ when calculating the final value of the phase margin. The value of z_f (and therefore also p_f) should be adjusted by trial and error until maximum phase compensation is attained. If this proves to be inadequate, then more lag-lead compensation is applied, and subsequently the phantom zero compensation is recalculated. Finally, the step response of the closed loop amplifier is checked to see if it meets specifications.

Example 7.6: Assume that we have an amplifier which has loop-gain poles at $p_1 = 100$ kHz, $p_2 = 200$ kHz, and $p_3 = 400$ kHz and no loop-gain zeros. The low-frequency value of the return ratio is 500, and the low-frequency asymptotic gain is 10. The feedback resistor R_f has a value of 10 kΩ and $R_s/R_i' = 1$. Compensate the amplifier so that its step response has an overshoot of 4.3% or less.

As a first step, we evaluate the phase at the loop-gain crossover frequency by examining

$$T(jf) = \frac{500}{(1 + jf/100)(1 + jf/200)(1 + jf/400)}$$

It is found that the loop-gain crossover occurs at 1.565 MHz, and the loop-gain phase attained at that frequency is $-244.72°$. This amplifier is unstable so that it definitely needs compensation.

To find out how much phase correction can be obtained by using phantom zero compensation, we substitute into (7.70) to find that

$$p_f = 11z_f$$

and the maximum phase correction is found from (7.84)

$$\phi_{\max} = 56.4°$$

Using (7.85) we find that we need lag-lead compensation to bring the phase at crossover to a value of approximately $-160°$ before phantom zero compensation can be considered. After a small amount of trial and error work, it is found that if the pole p_1 is moved to $p_1' = 3.33$ kHz and pole p_2 is moved to $p_2' = 6$ MHz (which in lag-lead compensation terms means that the pole relocation factor $\gamma = 30$), then an evaluation of

$$T(jf) = \frac{500}{(1 + jf/3.33)(1 + jf/400)(1 + jf/6000)}$$

shows that at the crossover frequency of 736.6 kHz, the loop-gain phase is $-159°$.

The two lowest poles of 3.33 kHz and 400 kHz and the return ratio of 500 are used to find that for phantom zero compensation we need $z_f = 542$ kHz and $p_f = 11(542)$ kHz. Those (initial) values are used to perform some additional numerical exploration to find locations of z_f and p_f that produce the desired phase margin. After a few calculations it is found that with $z_f = 640$ kHz and $p_f = 11(640)$ kHz, the expression

$$T(jf) = \frac{500(1 + jf/640)}{(1 + jf/3.33)(1 + jf/400)(1 + jf/6000)(1 + jf/7040)}$$

shows that at the loop-gain crossover frequency of 1.11 MHz the phase is $-119.4°$ for a phase margin of 60.6°. This is not the 65° phase margin sought, but this was never a second-order system, so it is possible that this phase margin is adequate to produce a reasonable step response. A SPICE simulation of the final data, which was arranged in the manner of Fig. 7.12, shows that this amplifier has a relatively conservative closed loop step response with an overshoot of 0.1%. ■

EXERCISES

7.1. Some approximations were used to obtain the data for Table 7.1. We want to see which makes the amplifier appear more unstable, Table 7.1 or the equations from which the data for the tables was obtained.

A two-pole feedback amplifier has the parameters $T_0 = 3$, $p_1 = 10$ krad/s, and $p_2 = 30$ krad/s. Find ζ, ω_c, ϕ_m, and ω_3 of this amplifier from Table 7.1 and again directly from the equations. Compare the values. Is (7.19) satisfied?

7.2. Repeat Prob. 7.1 for an amplifier with $T_0 = 50$, $p_1 = 10$ krad/s, and $p_2 = 500$ krad/s.

7.3. What is the effect on the loop-gain 3-dB frequency and on the closed loop-gain 3-dB frequency of the amplifier of Example 7.1, if the capacitor used for compensation is twice the calculated value?

7.4. A feedback amplifier with a second-order Butterworth response is desired. At the outset the loop-gain poles are at $p_1 = 1$ Mrad/s and $p_2 = 10$ Mrad/s. $T_0 = 100$ but it is much greater than the needed loop gain of 50.

(a) Obtain as much compensation as possible using the compensated attenuator method.

(b) The performance obtained is inadequate, so now apply narrowbanding to meet the specification. Calculate the loop-gain 3-dB frequency and the closed loop-gain 3-dB frequency.

(c) Instead of the narrowbanding in part (*b*), use lag-lead compensation to meet the specification. Calculate the loop-gain 3-dB frequency and the closed loop-gain 3-dB frequency.

7.5. A feedback amplifier with a second-order Butterworth response is desired. At the outset the loop-gain poles are at $p_1 = 1$ Mrad/s and $p_2 = 10$ Mrad/s, and $T_0 = 100$.

(a) Compensate using *only* narrowbanding. Calculate the loop-gain 3-dB frequency and the closed loop-gain 3-dB frequency. Compare performance to that obtained in Prob. 7.4b.

(b) Compensate using *only* lag-lead compensation. Calculate the loop-gain 3-dB frequency and the closed loop-gain 3-dB frequency. Compare performance to that obtained in Prob. 7.4c.

(c) The stage responsible for p_1 has at its input a resistance $R_1 = 20$ kΩ and a capacitance $C_1 = 50$ pF. Find the value of the compensating capacitance needed to obtain the compensation of part (a).

(d) Find the values of the compensating elements needed to obtain the compensation of part (b).

7.6. If the capacitor C_c which is used in the compensated attenuator is too large, then the attenuator is said to be overcompensated. If the converse is true it is said to be undercompensated. In Example 7.4:

(a) Write the expression for return ratio before compensation is applied and find the phase margin.

(b) Write the expression for the return ratio when the attenuator is perfectly compensated and find the phase margin.

(c) Assume that the attenuator is overcompensated with $C_c = 66.7$ pF. Working with Fig. 7.10, find the expression for $T(j\omega)$ and find the phase margin for this case. (Now $T(j\omega)$ will have one zero and three poles.)

(d) Assume that the attenuator is undercompensated with $C_c = 16.7$ pF. Again use Fig. 7.10 to find the expression for $T(j\omega)$ and find the phase margin for this case. (Here too $T(j\omega)$ will have one zero and three poles.)

7.7. A two-pole feedback amplifier has poles at $p_1 = 50$ kHz and $p_2 = 100$ kHz. The low-frequency loop gain $T(0) = 100$. The low-frequency asymptotic gain is $G_\infty(0) = 50$ and $R_i' = 10R_s$. The feedback resistor R_f has a value of 20 kΩ.

(a) Use phantom zero compensation to obtain a second-order Butterworth response. Find the location of the loop-gain zero z_f and the loop-gain pole p_f.

(b) Plot the magnitude and angle of $T(jf)$ as a function of frequency and find the loop-gain crossover frequency and the phase margin. Note how the zero z_f causes an upward swing in the phase characteristic in the vicinity of loop-gain crossover.

(c) By how many degrees does the pole p_f affect the phase margin?

7.8. A two-transistor amplifier with $T_0 = 120$ has only two significant poles. They are $p_1 = 0.15$ MHz and $p_2 = 2$ MHz. A second-order Butterworth response is desired for the closed loop amplifier. For the following three methods of compensation, compare the loop-gain bandwidth and the closed loop bandwidth.

(a) Narrowbanding.

(b) Lag-lead compensation. If $R_1 = 1$ kΩ find C_1 and the components needed for compensation.

(c) Phantom zero compensation. If $R_f = 10$ kΩ, what is the value of C_f needed for compensation? If the low-frequency asymptotic gain is $G_\infty(0) = 20$ and $R_i' = 2R_s$, find the influence the new pole has on the phase margin in this case.

7.9. It is desired to verify all the calculations made in Example 7.6.

(a) Find the loop-gain crossover frequency and the phase attained at that frequency.

(b) Find the relationship between the zero z_f and the pole p_f and find the maximum value of phase correction that this combination can provide.
(c) Relocate the pole p_1 downward by a factor γ, and the pole p_2 upward by a factor γ, choosing γ so that phantom zero compensation has a chance to succeed.
(d) Carry out phantom zero compensation and select a value of C_f.
(e) Write the resultant expression for $T(jf)$ and find the phase margin.
(f) Draw a diagram similar to Fig. 7.12 which will allow verification of the design by using a circuit analysis program. If a circuit analysis program is available, evaluate the closed loop step response.

8

Feedback Amplifier Sensitivity

IT WAS LEARNED IN CHAP. 1 that the reason feedback is used in amplifiers is to obtain good performance from inferior equipment, the price being an increase in the quantity of equipment needed to obtain a desired performance specification. It was also shown that the addition of feedback to an amplifier reduces its sensitivity with respect to gain variation of the active elements. It also reduces the distortion of the feedback amplifier by a factor corresponding to the return difference of the feedback amplifier. The latter statement is exact, whereas the results pertaining to sensitivity with respect to gain variation were based on the assumption that the variations in the gain of the active amplifier elements is very small. A more accurate equation for sensitivity will now be derived and methods will then be presented for specifying the loop gain required for meeting a desired design objective.

8.1 *Some Definitions*

Before we proceed we need to introduce the following notation

$$T_n \equiv \text{nominal value of } T \tag{8.1a}$$

$$T_h \equiv \text{high value of } T \tag{8.1b}$$

$$T_l \equiv \text{low value of } T \tag{8.1c}$$

The loop-gain fractional variations are defined as

$$\delta \equiv \frac{(\text{change in } T)}{T} \tag{8.2a}$$

$$\delta_+ \equiv \frac{T_h - T_n}{T_n} \tag{8.2b}$$

$$\delta_- \equiv \frac{T_l - T_n}{T_n} \tag{8.2c}$$

In order to get a feeling for the above definitions, we shall define an ideal i-stage hypothetical amplifier. Assume it is possible to connect i common emitter transistors in cascade, (ignoring the fact that this cannot be done for even i), and to apply feedback by connecting a resistor from the collector of the last transistor to the base of the first transistor. Then it is clear that T is proportional to β^i, so that $T = k\beta^i$, where k is a proportionality constant. The transistor current gain β can take on the nominal value β_n, the high value β_h, and the low value β_l. The high and low values of return ratio for this amplifier are related to the nominal value by

$$T_h = \left(\frac{\beta_h}{\beta_n}\right)^i T_n \tag{8.3a}$$

$$T_l = \left(\frac{\beta_l}{\beta_n}\right)^i T_n \tag{8.3b}$$

The fractional variations in return ratio for this hypothetical amplifier are

$$\delta_+ = \left(\frac{\beta_h}{\beta_n}\right)^i - 1 \tag{8.4a}$$

$$\delta_- = \left(\frac{\beta_l}{\beta_n}\right)^i - 1 \tag{8.4b}$$

Care must be exercised not to assume that for all i-transistor amplifiers the loop gain will be proportional to beta to the ith power. It depends entirely on the configuration of the amplifier. In a three-transistor shunt–shunt connection the loop gain will depend on β^3, but for a three-transistor series–series amplifier the loop gain depends directly on the beta of the second transistor, but not as strongly on the betas of the first and third transistors. To assume that for an n-transistor amplifier the return ratio depends on β^n does no harm in a preliminary analysis, since this takes the most pessimistic point of view. The dependence of T on beta should eventually be established for the amplifier configuration under consideration, and some of the exercises at the end of this chapter address this question.

It will be assumed in this analysis that in the asymptotic gain formula (2.18), the last term makes an insignificant contribution, so we shall use

$$G_f = \frac{G_\infty T}{1 + T} \tag{8.5}$$

to relate loop gain T to the closed loop gain G_f. Corresponding to T_n, T_h, and T_l, we have

$$G_{fn} \equiv \text{nominal value of } G_f \tag{8.6a}$$

$$G_{fh} \equiv \text{high value of } G_f \tag{8.6b}$$

$$G_{fl} \equiv \text{low value of } G_f \tag{8.6c}$$

The feedback amplifier closed loop-gain variations are defined as

$$\Delta \equiv \frac{(\text{change in } G_f)}{G_f} \tag{8.7a}$$

$$\Delta_+ \equiv \frac{G_{fh} - G_{fn}}{G_{fn}} \tag{8.7b}$$

$$\Delta_- \equiv \frac{G_{fl} - G_{fn}}{G_{fn}} \tag{8.7c}$$

8.2 *Sensitivity with Respect to Loop-Gain Variations*

The sensitivity S_T of the closed loop amplifier gain G_f with respect to variations in the loop gain T, is defined as the fractional change in the closed loop amplifier gain divided by the fractional change in the loop gain

$$S_T = \frac{(\text{change in } G_f)/G_f}{(\text{change in } T)/T} \tag{8.8}$$

If the loop gain T changes by a fraction δ then the closed loop gain G_f will change by a fraction Δ. When the two are related by (8.5) we obtain

$$G_f(1 + \Delta) = \frac{G_\infty T(1 + \delta)}{1 + T(1 + \delta)} \tag{8.9}$$

Subtracting (8.5) from (8.9) we obtain

$$G_f \Delta = \frac{G_\infty T\delta}{(1 + T)[1 + T(1 + \delta)]} \tag{8.10}$$

This is divided by (8.5), with the result

$$\Delta = \frac{\delta}{1 + T(1 + \delta)} \tag{8.11}$$

From (8.11) and (8.8) we see that the sensitivity of the amplifier closed loop gain with respect to loop-gain variations is

$$S_T = \frac{\Delta}{\delta} = \frac{1}{1 + T(1 + \delta)} \tag{8.12}$$

We note that if the loop-gain variations are small, then (8.12) reduces to

$$S_T = \frac{1}{1 + T} \quad \text{for } \delta \ll 1 \tag{8.13}$$

which is in agreement with (1.15), which was derived for differentially small changes in the loop gain T.

Example 8.1: Let $G_\infty = 11$. Suppose the nominal value of T is 10 and it varies by $\pm 20\%$. Find the nominal value, and the high and low values of the closed loop gain G_f.

From the specification we have

$$T_n = 10 \qquad T_h = 12 \qquad T_l = 8$$

Using (8.5) we find

$$G_{fn} = 10 \qquad G_{fh} = 10.1538 \qquad G_{fl} = 9.7778$$

From this data we find that

$$\Delta_+ = 0.0154 = 1.54\% \qquad \Delta_- = -0.0222 = -2.22\%$$

It becomes apparent that for equal upward and downward variations in T, the variations in G_f turn out to be unequal, the downward variation being larger than the upward variation. ■

8.3 *Choosing Loop Gain to Meet Specifications Approximately*

We are now in a position to specify the required loop gain T, to meet a given sensitivity objective. From (8.11) we obtain

$$T = \frac{(\delta/\Delta) - 1}{1 + \delta} \tag{8.14}$$

Example 8.2: Consider a three-transistor shunt–shunt feedback amplifier, with the three common emitter transistors connected in cascade. The current gain β of each transistor can vary from the nominal value β_n by

$\pm 20\%$. It is desired that closed loop amplifier gain vary no more than $\pm 1\%$ from design center value.

The amplifier loop gain is proportional to β^3. Therefore the loop-gain nominal value T_n is related to the nominal value of transistor current gain β_n, through some arbitrary constant k:

$$T_n = k\beta_n^3$$

When β takes on its high value, the fractional change in return ratio is

$$\delta_+ = \frac{k[\beta_n(1+0.2)]^3 - k\beta_n^3}{k\beta_n^3} = 0.728$$

and when β is low it is

$$\delta_- = \frac{k[\beta_n(1-0.2)]^3 - k\beta_n^3}{k\beta_n^3} = -0.488$$

From the closed loop-gain specification we have

$$\Delta_+ = 0.01$$

and

$$\Delta_- = -0.01$$

Substituting the data for Δ_- and δ_- into (8.14) we obtain

$$T = 93.4$$

This is the required nominal value of T when the current gain β is at its nominal value of β_n. That value of T is needed to meet the specification for Δ_- and δ_-. Again using (8.14), we find that the required nominal T needed to meet the specification for Δ_+ and δ_+ is

$$T = 41.6$$

■

We see from the last example that the computed nominal value for T when we are dealing with upward variations is inadequate if β varies downward, and the opposite is true for the reverse situation. There must be a value of nominal T which lies somewhere between 41.6 and 93.4, which could meet the specification if it were restated properly.

8.4 *Choosing Loop Gain to Satisfy Specifications Exactly*

If we wish to obtain a nominal value for T that will keep variations in G_f within specifications, we must find an expression that takes all the specified data into consideration. To do this, only the location of the extreme points of operation will be considered. The question of where the nominal point of operation ends up will not be of interest right now. Previously, we considered all variations of G_f to be computed relative to a nominal value G_{fn}, and we saw in Example 8.1 that G_{fn} was tied to T_n through (8.5). Now all variations in G_f will be taken relative to G'_{fn}, whose value is not known at the outset.

When T takes on its largest value, we expect G_f to do the same. Equation (8.9) must now be recast into the two relationships

$$G_{fh} = G'_{fn}(1 + \Delta_+) = \frac{G_\infty T_n(1 + \delta_+)}{1 + T_n(1 + \delta_+)} \tag{8.15}$$

$$G_{fl} = G'_{fn}(1 + \Delta_-) = \frac{G_\infty T_n(1 + \delta_-)}{1 + T_n(1 + \delta_-)} \tag{8.16}$$

Dividing (8.15) by (8.16) and solving for T_n, we obtain

$$T_n = \frac{1}{\Delta_+ - \Delta_-}\left[\frac{1 + \Delta_-}{1 + \delta_-} - \frac{1 + \Delta_+}{1 + \delta_+}\right] \tag{8.17}$$

This is a very useful expression for finding the nominal value of T to meet a high and low specification, simultaneously. To find G'_{fn} from G_{fh} or G_{fl}, we get from (8.15) and (8.16)

$$G'_{fn} = \frac{G_{fh}}{1 + \Delta_+} = \frac{G_{fl}}{1 + \Delta_-} \tag{8.18}$$

An application will be demonstrated in the following example.

Example 8.3: It was found in Example 8.2 that we had an amplifier with $\delta_+ = 0.728$, $\delta_- = -0.488$, $\Delta_+ = 0.01$, $\Delta_- = -0.01$. In addition let $G_\infty = 11$. It will be interesting to see if the specification that (8.17) produces for T_n is much different from the result obtained in Example 8.2.

Applying (8.17) to the data we obtain

$$T_n = \frac{1}{0.01 - (-0.01)}\left[\frac{1 + (-0.01)}{1 + (-0.488)} - \frac{1 + 0.01}{1 + 0.728}\right] = 67.46$$

which is somewhere between the two values of T found in Example 8.2. We now list the correspondence between the various values of T and G_f, which are related through (8.5).

$$T_l = 34.54 \qquad G_{fl} = 10.69$$

$$T_n = 67.46 \qquad G_{fn} = 10.84$$

$$T_h = 116.6 \qquad G_{fh} = 10.91$$

Using (8.18) we find that

$$G'_{fn} = \frac{G_{fh}}{1 + \Delta_+} = \frac{10.91}{1.01} = 10.80$$

and this value of closed loop gain does not equal G_{fn}, and it was not meant to do so. It is merely a more reasonable basis of reference for the closed loop amplifier gain variations than is G_{fn}.

We can finally conclude that we can obtain the nominal closed loop amplifier gain $G'_{fn} = 10.80$ which will vary within $\pm 1\%$ when the nominal value of loop-gain is set at 67.46 and is allowed to vary $+72.8\%$ and -48.8%. ■

8.5 *Impedance Specification*

There are situations in which an amplifier voltage-gain variation is specified that must not be exceeded as the load impedance is varied over a range of values. It might be stated, for example, that the output voltage of the amplifier may not vary more than $\pm 10\%$ as the load is varied from an open circuit down to 100 Ω. We need to find the value of output impedance that will allow us to meet this specification.

The Thevenin equivalent circuit at the output terminals of the amplifier is shown in Fig. 8.1 for two load impedances.

It is assumed that $Z_{L2} > Z_{L1}$, hence $V_2 > V_1$. We shall define as in (8.7), using V_{ref} as an output reference voltage

$$\Delta_{L+} \equiv \frac{V_2 - V_{\text{ref}}}{V_{\text{ref}}} \tag{8.19}$$

and

$$\Delta_{L-} \equiv \frac{V_1 - V_{\text{ref}}}{V_{\text{ref}}} \tag{8.20}$$

The output voltages for the two situations depicted in Fig. 8.1 are given by

$$V_2 = V_{\text{ref}}(1 + \Delta_{L+}) = \frac{Z_{L2}}{Z_0 + Z_{L2}} E_0 \tag{8.21}$$

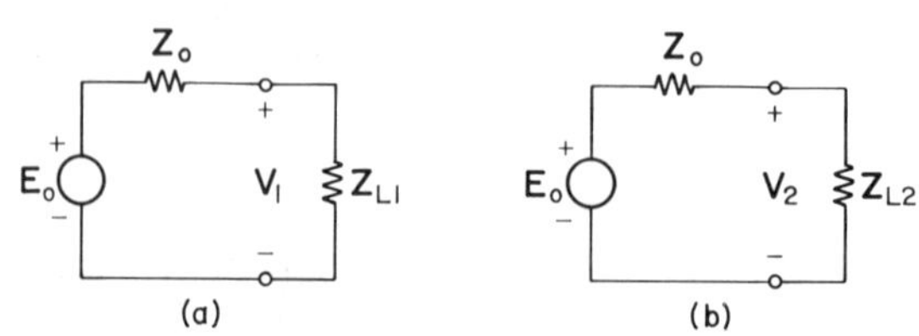

Figure 8.1 Thevenin equivalents of an amplifier.

and

$$V_1 = V_{\text{ref}}(1 + \Delta_{L-}) = \frac{Z_{L1}}{Z_0 + Z_{L1}} E_0 \tag{8.22}$$

Taking the ratio of the last two equations and solving for the required output impedance, we find

$$Z_0 = \frac{\Delta_{L+} - \Delta_{L-}}{\dfrac{1 + \Delta_{L-}}{Z_{L1}} - \dfrac{1 + \Delta_{L+}}{Z_{L2}}} \tag{8.23}$$

Example 8.4: It is specified that for an amplifier, the load voltage may not vary more than +10% and −5% from some imaginary reference value as the load impedance varies from 5000–100 Ω. What should be the amplifier output impedance?

From the data it is clear that $Z_2 = 5000\ \Omega$, $Z_1 = 100\ \Omega$, $\Delta_{L+} = 0.1$, and $\Delta_{L-} = -0.05$. This is substituted into (8.23) to produce the result

$$Z_0 = \frac{0.1 - (-0.05)}{(0.95/100) - (1.05/5000)} = 16.15\ \Omega$$

If the output impedance is kept below this value, then the output voltage variation will fall within the specified range. ■

8.6 *Conclusion*

We have shown that it is possible to arrive at accurate values of return ratio from a stated set of specifications. If the specifications pertain to loop gain and closed-loop gain variations, then we proceed to use (8.17) in a very straightforward manner. If there are additional specifications pertaining to input and output impedance variation, then (8.23) can be used to determine the required feedback amplifier output impedance. In case there are further constraints imposed on the feedback amplifier design, such as requirements on distortion and power-supply ripple suppression, then the material of this chapter can serve as a good starting point, but further progress must be made by using trial and error techniques.

EXERCISES

8.1. For the series–shunt two-transistor feedback amplifier of Example 2.1, which is reproduced in Fig. P8.1, it was found that $T_n = 34.9$. Both transistors have current gains of 100, with a tolerance of −50% and

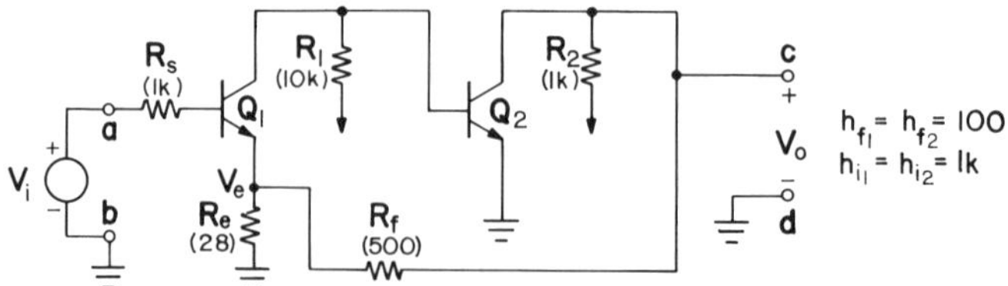

Figure P8.1

+100%. Find T_h and T_l and compare them to the hypothetical ideal amplifier of Sec. 8.1.

8.2. For the shunt–series two-transistor feedback amplifier of Prob. 2.3, which is reproduced in Fig. P8.2, both transistors have current gains of 100, with a tolerance of −50% and +100%. Find T_h and T_l and compare them to the hypothetical ideal amplifier of Sec. 8.1.

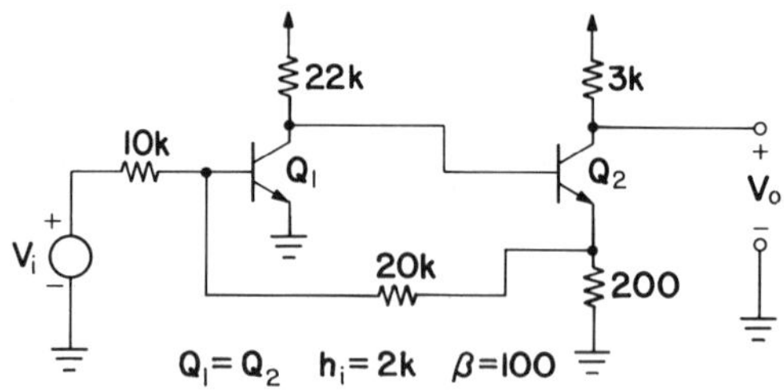

Figure P8.2

8.3. For the series–series three-transistor feedback amplifier of Prob. 2.5, which is reproduced in Fig. P8.3, all transistors have current gains of 100, with a tolerance of −50% and +100%. Find T_h and T_l and compare them to the hypothetical ideal amplifier of Sec. 8.1.

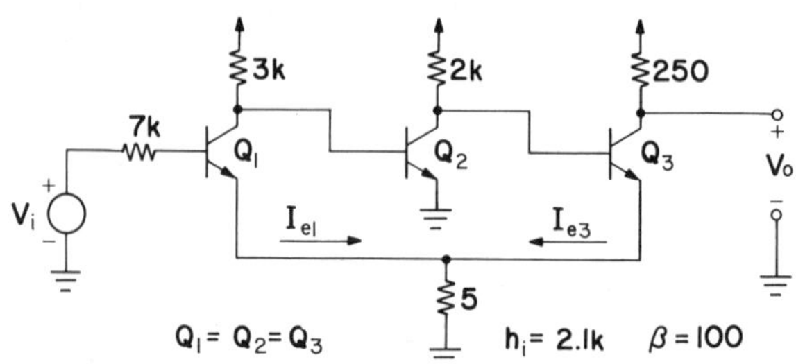

Figure P8.3

8.4. We saw in Example 8.1 that equal up–down variations in T do not produce equal up–down variations in G_f.

(a) Find a new nominal reference value G'_{fn}, which is the mean of G_{fh} and G_{fl}. Verify that Δ'_+ and Δ'_- are equal in magnitude, when the closed loop gain departs from G'_{fn}.

(b) Find a new nominal loop-gain reference value T'_n which corresponds to G'_{fn}. Find δ'_+ and δ'_- relative to T'_n.

8.5. For a certain amplifier application, transistors are available, which may be represented to an adequate approximation by the following parameters at room temperature:

$$h_i = \text{input impedance} = 2k$$

$$\beta = \text{current gain} = 50$$

As temperature is varied from the nominal value to the operating limits, it may be assumed that the only change in transistor parameters is a change in β by $+100\%$ and -50%. To assure adequate collector–emitter voltage, and to limit dissipation to safe limits, the load resistor for each transistor must be chosen in the range 1–10$k\Omega$.

The amplifier is supplied by a source that may be represented by a generator of open-circuit voltage E_s in series with a resistor of 100 ohms. For a fixed operating temperature and with external load resistance in the range of 400–4000 ohms connected to the output of the amplifier, the voltage across the load shall not change by more than $\pm 5\%$ (for constant E_s) under the worst conditions of operation. The nominal gain G_f, which is the ratio of voltage across the load to E_s, is to be 20. For a fixed load impedance, as the temperature is varied throughout the operating range, G_f should not change by more than $\pm 10\%$.

Ignoring the effect of impedance associated with bias circuits, determine:

(a) An amplifier configuration that will be capable of meeting these requirements including the number of transistors necessary.

(b) The values of all circuit elements used in the amplifier.

9

Oscillators

When designing feedback amplifiers, every effort is made to ensure that the amplifier should be stable and never oscillate under any operating conditions. Oscillators are circuits, which are deliberately designed with enough positive feedback to cause instability. The circuit components are intentionally chosen so that the oscillator will output a sinusoidal wave at a very specific frequency and will continue doing that as long as power is applied to the circuit.

An oscillator can be viewed as a feedback amplifier which is deliberately made unstable. When the Nyquist plot for a feedback amplifier passes through the point $s = -1$, then it is an indication that the closed loop response has a pole on the $j\omega$ axis at some frequency ω_0, and is capable of maintaining a sustained sinusoidal oscillation at this frequency. We therefore use $T(j\omega_0) = -1$ as a *minimum* condition for oscillation.

A slight change in circuit parameters could cause the $j\omega$-axis pole to move into the left half of the s plane, causing oscillations to cease. For this reason, it is desirable in practical oscillators to have the Nyquist plot pass to the left of the point $s = -1$. This is equivalent to requiring the magnitude of T to exceed unity at the frequency of oscillation, so that most oscillators will, in fact, have poles slightly to the right of the $j\omega$ axis. So for practical oscillators, we require that

$$\operatorname{Re} T(j\omega_0) \leq -1 \tag{9.1a}$$

$$\operatorname{Im} T(j\omega_0) = 0 \tag{9.1b}$$

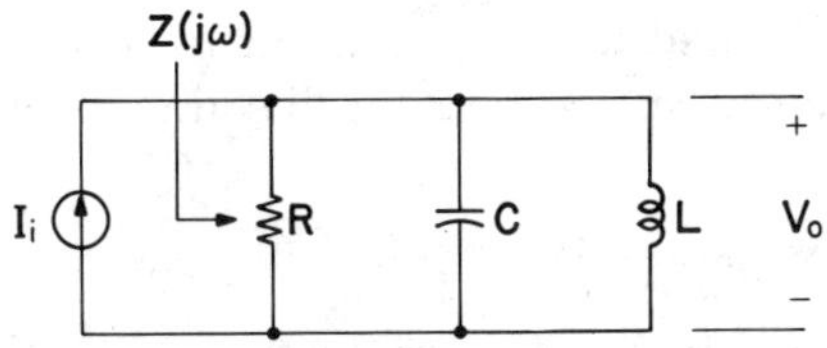

Figure 9.1 Parallel-tuned *RLC* circuit.

The oscillation buildup is eventually limited by the nonlinearities of the active elements, namely the active-device approaches saturation or cutoff or both. A frequency-selective circuit serves two purposes in feedback oscillators. Its first function is to ensure that the conditions of oscillation are met only at the desired frequency ω_0, and its second function is to remove the higher harmonics of the distorted output signal, so that a "clean" sinusoidal signal may be seen by the load. Before proceeding with the subject of this chapter, a short review of tuned circuits will be undertaken.

9.1 *Parallel RLC Circuits*

We shall analyze the parallel *RLC* circuit of Fig. 9.1, to see how the output voltage V_o varies as a function of frequency for a constant input current I_i.

The output voltage V_o is related to I_i through $Z(j\omega)$, which is given by

$$Z(j\omega) = \frac{1}{\dfrac{1}{R} + j\omega C + \dfrac{1}{j\omega L}} = \frac{R}{1 + j\omega CR + \dfrac{R}{j\omega L}} \tag{9.2}$$

Define

$$\omega_0 = 2\pi f_0 = 1/\sqrt{LC} \equiv \text{resonant frequency} \tag{9.3}$$

and

$$Q = \frac{R}{\omega_0 L} = \omega_0 RC \equiv \text{quality factor} \tag{9.4}$$

Then (9.2) can be written in the very convenient form

$$Z(j\omega) = \frac{R}{1 + jQ\left[\dfrac{f}{f_0} - \dfrac{f_0}{f}\right]} \tag{9.5}$$

It is clear from (9.5) that the impedance of the parallel-tuned circuit is highest at resonance, when it is equal to the parallel resistance R. The impedance decreases as the operating frequency is increased or decreased away from resonance. To find the 3-dB frequencies of $Z(j\omega)$, we equate the imaginary term in the denominator to +1 and solve the resulting quadratic equation. Only the positive root is retained and it is found that the upper 3-dB frequency f_H is

$$\frac{f_H}{f_0} = \frac{1}{2Q} + \sqrt{1 + \left(\frac{1}{2Q}\right)^2} \tag{9.6}$$

When the imaginary part of the denominator is set to -1 and again only the positive root is retained in the resultant quadratic equation, it is found that

$$\frac{f_L}{f_0} = -\frac{1}{2Q} + \sqrt{1 + \left(\frac{1}{2Q}\right)^2} \tag{9.7}$$

The difference between the frequencies f_H and f_L is the 3-dB bandwidth of the tuned circuit and is given by

$$BW = \frac{f_0}{Q}\ (\mathrm{Hz}) = \frac{\omega_0}{Q}\ (\mathrm{rad/s}) \tag{9.8}$$

Now it becomes apparent why Q is referred to as the quality factor of the resonant circuit. The higher the value of Q, the narrower is the 3-dB bandwidth of the resonant circuit, the better its frequency selectivity. It must be kept in mind that the value of R used in all the above relationships must account for all resistive losses in the parallel resonant circuit.

9.2 *Parallel-Tuned Circuits with Series Loss*

Before we can include all resistive losses in the calculation of the Q of the resonant circuit, we have to consider the fact that usually the inductor used in resonant circuits has a series resistance which represents the coil losses, as shown in Fig. 9.2*a*. It would be useful to convert this circuit to the parallel form shown in Fig. 9.2*b*, so that the expressions derived in Sec. 9.1 can be applied directly.

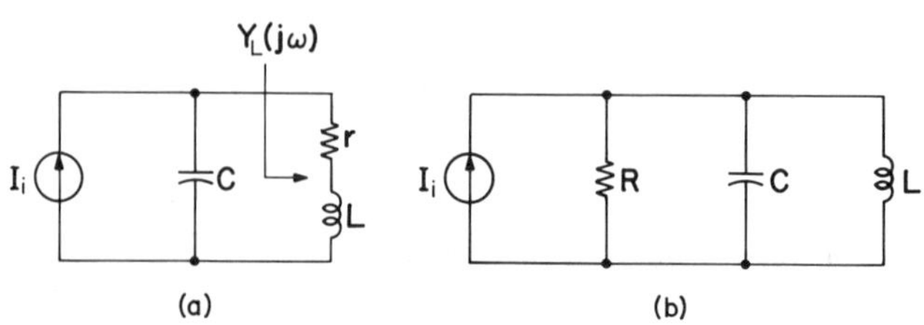

Figure 9.2 (*a*) Inductor with series loss. (*b*) Desired equivalent.

To find the equivalent circuit of the coil in a narrow band of frequencies in the vicinity of ω_0, we write the admittance of the coil in Fig. 9.2*a* at the resonant frequency ω_0

$$Y_L(j\omega_0) = \frac{1}{r + j\omega_0 L} = \frac{r - j\omega_0 L}{r^2 + \omega_0^2 L^2} \tag{9.9}$$

If the series resistance is much smaller than the inductive reactance, namely

$$r \ll \omega_0 L \tag{9.10}$$

then (9.9) reduces to

$$Y_L(j\omega_0) \approx \frac{r}{\omega_0^2 L^2} + \frac{1}{j\omega_0 L} \tag{9.11}$$

We see that for frequencies which are very close to resonance, we can use the parallel equivalent circuit of Fig. 9.2*b*, in which

$$R = \frac{\omega_0^2 L^2}{r} \tag{9.12}$$

and the value of the coil is left unchanged.

Using this value of R in (9.4), we find that Q can be expressed in terms of the parameters of the series coil

$$Q = \frac{\omega_0 L}{r} \tag{9.13}$$

This is the Q of the inductor in which a series resistor is used to characterize the coil losses, and now we see that condition (9.10), under which the transformation applies, is equivalent to saying that Q must be much greater than 1. Another form used for finding the shunt resistance R from the series loss resistance r is obtained by substituting (9.13) into (9.12) resulting in

$$R = Q^2 r \tag{9.14}$$

Example 9.1: A coil has an inductance of 100 mH and a series resistance of 25 Ω. The coil is in parallel with a 0.1 μF capacitor and also in parallel with a 120 kΩ resistor. Find the parameters of the circuit near resonance.

We will assume at the outset that the Q of this resonant circuit is high and later verify that this is a valid assumption. The resonant frequency is found using (9.3), hence $\omega_0 = 10$ krad/s. The Q of the coil (not including the effect of the 120 kΩ shunting resistor) is found using (9.13), hence $Q_L = 40$. Using (9.14) we find that the equivalent parallel resistance for the coil is 40 kΩ. This is now combined with the parallel

120 kΩ resistor, for a total shunting value of 30 kΩ. The equivalent circuit of the form of Fig. 9.2*b* will contain a 100 mH coil, a 0.1 μF capacitor, and a shunt resistance of 30 kΩ. The Q of this equivalent circuit can be calculated using (9.4), so that $Q = 30$ and it is substantially greater than 1. The bandwidth of this tuned circuit is found from (9.8) with the result $BW = 333.3$ rad/s. ■

9.3 *Transformerlike Resonant Circuits*

In this section we consider parallel resonant circuits in which the lossy element appears across a portion of the inductor or capacitor as shown in Fig. 9.3*a*. The elements Z_1 and Z_2 represent either capacitive elements and Z_3 then represents an inductive element, or Z_1 and Z_2 can represent a tapped coil, in which case, Z_3 would represent a capacitive element. Some equivalent circuits for transformer coupled coils will also be derived.

Consider the parallel-tuned circuit of Fig. 9.3a. We wish to find a resistance R_{eq}, shown in Fig. 9.3*b*, which can replace the resistances R_i and R_o, leaving the power losses in the circuit unchanged. Assuming the resistors load the circuit only slightly (which is tantamount to saying that the Q of the circuit is very high), then the reactive elements determine the voltages V_1 and V_2, and are given by

$$V_1 = \frac{Z_1}{Z_1 + Z_2} V \tag{9.15}$$

$$V_2 = \frac{Z_2}{Z_1 + Z_2} V \tag{9.16}$$

Equating the power lost in Figs. 9.3*a* and 9.3*b* we conclude that

$$R_{\text{eq}} = \frac{(Z_1 + Z_2)^2}{\dfrac{Z_1^2}{R_o} + \dfrac{Z_2^2}{R_i}} \tag{9.17}$$

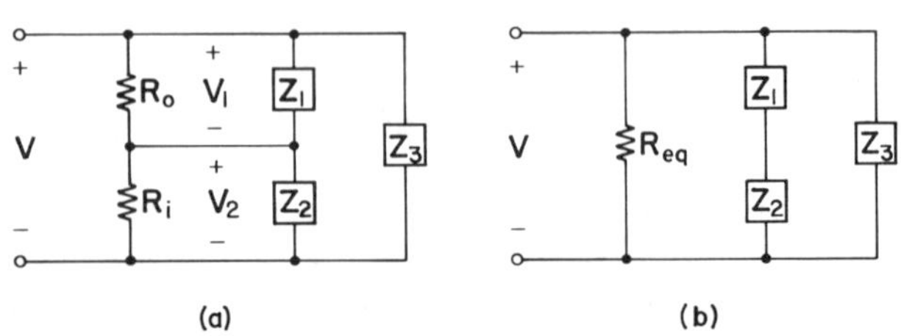

Figure 9.3 Parallel-tuned circuit with light loading.

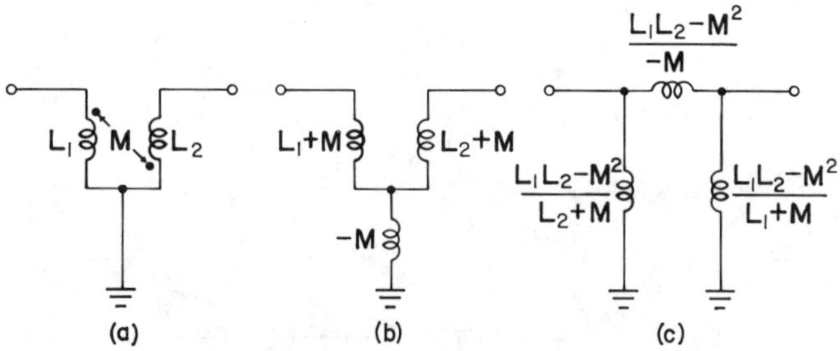

Figure 9.4 (*a*) Transformer. (*b*) Mesh impedance matrix equivalent. (*c*) PI equivalent.

We make the observations that if R_i and R_o are of the same order of magnitude, then

$$R_{\text{eq}} \approx R_i \quad \text{if } Z_1 \ll Z_2 \tag{9.18}$$

$$R_{\text{eq}} \approx R_o \quad \text{if } Z_1 \gg Z_2 \tag{9.19}$$

Another useful transformation is needed before we get into the subject of *LC*-tuned oscillators. We wish to find the PI equivalent of the transformer shown in Fig. 9.4*a*.

We assign voltages and currents to the two port, and write the mesh equations. We find that the same mesh equations can be obtained for Figs. 9.4*a* and 9.4*b*, so that these two circuits are equivalent. We use the *Y*-delta (or star-PI) transformation (which can be found in any elementary circuit theory text) to go from Figs. 9.4*b* to 9.4*c*, to complete the solution.

9.4 *The Colpitts and Hartley Oscillators*

In Fig. 9.5*a* we have a very general circuit, which represents either a Colpitts or a Hartley oscillator. The resistance R_L is included to take into consideration the load on the oscillator output, and R_b is included to account for the biasing resistors for the active device. In the equivalent circuit of Fig. 9.5*b*, the resistance R_i is the parallel combination of the transistor (or FET) input resistance and the resistances of the biasing circuitry. The resistance R_o is the parallel combination of the load resistance R_L and the output resistance of the transistor (or FET). The impedances Z_1, Z_2, and Z_3 are purely reactive. Figure 9.5*b* is the equivalent circuit which will be used for the calculation of return ratio.

We replace the dependent source $g_m V$ by an independent source of value g_m and calculate V. The return ratio is the negative of the returned

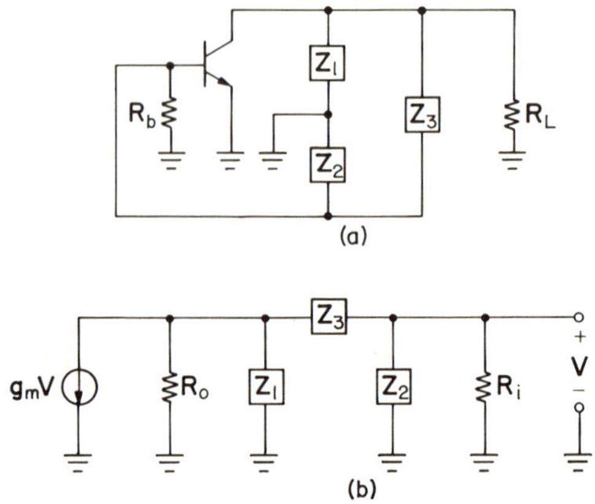

Figure 9.5 (*a*) Colpitts or Hartley oscillator. (*b*) Equivalent circuit.

voltage V, so after a little simplification it is found that $T(j\omega)$ is given by

$$T(j\omega) = \frac{g_m R_i R_o Z_1 Z_2}{R_i Z_1(Z_2 + Z_3) + R_o Z_2(Z_1 + Z_3) + R_i R_o(Z_1 + Z_2 + Z_3) + Z_1 Z_2 Z_3} \tag{9.20}$$

Since Z_1, Z_2, and Z_3 are purely reactive, then the numerator and the first two terms in the denominator are real. The last two denominator terms are imaginary. To meet the condition of oscillation of (9.1b) the imaginary part of (9.20) must vanish. We therefore require

$$R_i R_o(Z_1 + Z_2 + Z_3) + Z_1 Z_2 Z_3 = 0 \tag{9.21}$$

and it will be found later that the solution to this equation will yield the frequency of operation ω_0 of the oscillator.

We shall assume at this point that the term $Z_1 Z_2 Z_3$ does not affect the solution sufficiently to be needed in the above equation. Later it will be shown what conditions the circuit parameters must meet for this to be true. Therefore the following approximate condition will henceforth be used

$$Z_1 + Z_2 + Z_3 = 0 \tag{9.22}$$

to find ω_0. Having eliminated the last two terms in the denominator of (9.20), we substitute (9.22) into the remaining terms, and impose condition of oscillation (9.1a) on the result, to obtain

$$T(j\omega_0) = \frac{-g_m R_i R_o Z_1 Z_2}{R_i Z_1^2 + R_o Z_2^2} \leq -1 \tag{9.23}$$

From the above it follows that the requirement for oscillation is

$$g_m R_o \geq \frac{Z_1}{Z_2} + \frac{R_o}{R_i} \frac{Z_2}{Z_1} \tag{9.24}$$

The left-hand side represents the voltage gain of the active device. It is best to have the left-hand side of the equation exceed the right-hand side by a small margin, otherwise the magnitude of T becomes much greater than unity and the active device has to go very far into the nonlinear region of operation to reduce the magnitude of T. This can cause a great deal of distortion, which the tuned circuit may be incapable of removing. In FET oscillators R_i is the gate bias resistor R_g, and it is so much bigger than R_o that Z_2 would have to be extremely large in comparison with Z_1 if the second term on the right were to be used to bring both sides of (9.24) into reasonable balance. It is more practical to use the first term on the right to achieve that objective, hence we shall choose $Z_1 \gg Z_2$, so that for all practical purposes (9.24) reduces to

$$g_m R_o \geq \frac{Z_1}{Z_2} \tag{9.25}$$

The same expression is used in bipolar-transistor oscillators, but for a different reason. In most bipolar-transistor oscillators, the resistance R_o loading the output is an order of magnitude greater than the resistance R_i loading the input side of the resonant circuit. In view of this and the results of (9.18) and (9.19), we see that for minimum loading (hence for higher Q) of the resonant circuit, we should choose to make $Z_1 \gg Z_2$, and then (9.24) reduces to (9.25) as it did for FETs.

For FETs we can substitute $g_m = \mu/r_d$ and $R_o = r_d \| R_L$, to obtain

$$\mu \geq \left[1 + \frac{r_d}{R_L}\right] \frac{Z_1}{Z_2} \tag{9.26}$$

For bipolar transistors we can substitute $g_m = \beta/h_{ie}$ and $R_o = R_L \| (1/h_{oe})$, to get the form

$$\beta \geq \frac{h_{ie}}{R_L \| (1/h_{oe})} \frac{Z_1}{Z_2} \tag{9.27}$$

The foregoing results will now be applied to the Colpitts oscillator of Fig. 9.6. The radio frequency choke (RFC) is used in place of a collector resistor so that the loading on the resonant circuit will be minimized in an effort to keep its Q value high.

From the diagram it is clear that

$$Z_1 = \frac{1}{j\omega C_1} \qquad Z_2 = \frac{1}{j\omega C_2} \qquad Z_3 = j\omega L \tag{9.28}$$

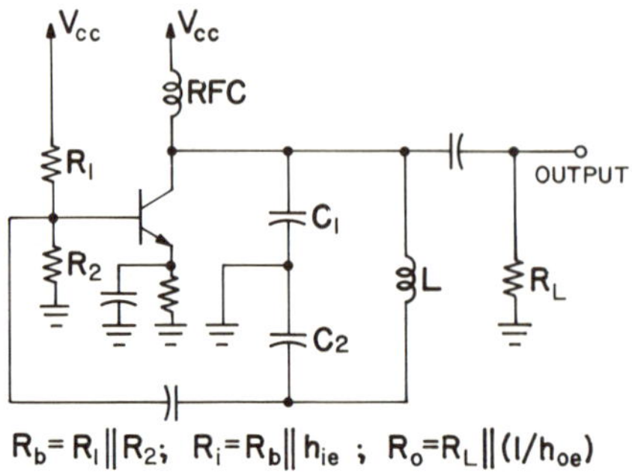

Figure 9.6 A practical example of a Colpitts oscillator.

We get upon substitution of these values into (9.22)

$$\omega_0^2 = \frac{1}{L}\left[\frac{1}{C_1} + \frac{1}{C_2}\right] \tag{9.29}$$

It becomes clear from the above expression that the inductor L resonates with the series combination of the capacitors C_1 and C_2. Had we substituted into (9.21) instead, we would have obtained the result

$$\omega_0^2 = \frac{1}{L}\left[\frac{1}{C_1} + \frac{1}{C_2}\right] + \frac{1}{R_i R_o C_1 C_2} \tag{9.30}$$

If the reactive elements of the Colpitts oscillator are chosen to satisfy

$$\frac{L}{C_1 + C_2} \ll R_i R_o \tag{9.31}$$

then the frequency of oscillation becomes independent of the resistive loading the resonant circuit, and the result produced by (9.22) is indistinguishable from that obtained from (9.21). Condition (9.31) is not difficult to obtain in practice.

When the values of (9.28) are substituted into (9.27), we find that the transistor voltage gain must satisfy

$$g_m R_o \geq \frac{C_2}{C_1} \tag{9.32}$$

It is clear from the above that we must choose $C_2 \gg C_1$. If the oscillator is to be tunable over a range of frequencies, then C_1 is a variable capacitor, since the series combination of the two capacitors is approximately C_1. The disadvantage in this method is that varying C_1 affects the ratio on the right-hand side of (9.32), which can cause variations in the output signal distortion as the frequency is changed. To remedy the situation a Hartley circuit is used if a tunable oscillator is desired.

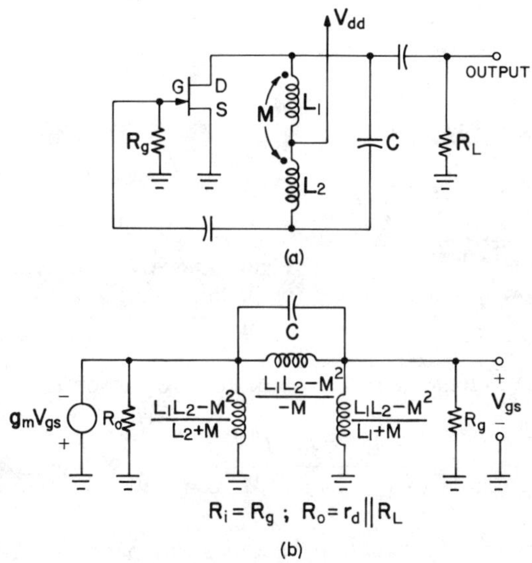

Figure 9.7 A practical example of a Hartley oscillator.

The Hartley oscillator is another example of a very common type of tuned circuit oscillator as shown in Fig. 9.7. It requires a tapped inductance, which is more difficult to obtain than the fixed inductance of the Colpitts oscillator, but as can be seen in Fig. 9.7, its resonant frequency is very readily varied by using a variable capacitor in place of C. The drain supply is introduced through the tap in the coil to dispense with the radio frequency choke that would otherwise be needed. The equivalent circuit was drawn using the transformations shown in Fig. 9.4. From the diagram it is clear that

$$Z_1 = j\omega\frac{L_1L_2 - M^2}{L_2 + M} \qquad Z_2 = j\omega\frac{L_1L_2 - M^2}{L_1 + M} \qquad Z_3 = \frac{L_1L_2 - M^2}{-M} \tag{9.33}$$

We get upon substitution of these values into (9.22)

$$\omega_0^2 = \frac{1}{(L_1 + L_2 + 2M)C} \tag{9.34}$$

It becomes clear from the above expression that the entire inductor resonates with the capacitor C. Had we substituted into (9.21) instead, we would have obtained the result

$$\omega_0^2 = \frac{1}{(L_1 + L_2 + 2M)C + \dfrac{L_1L_2 - M^2}{R_oR_i}} \tag{9.35}$$

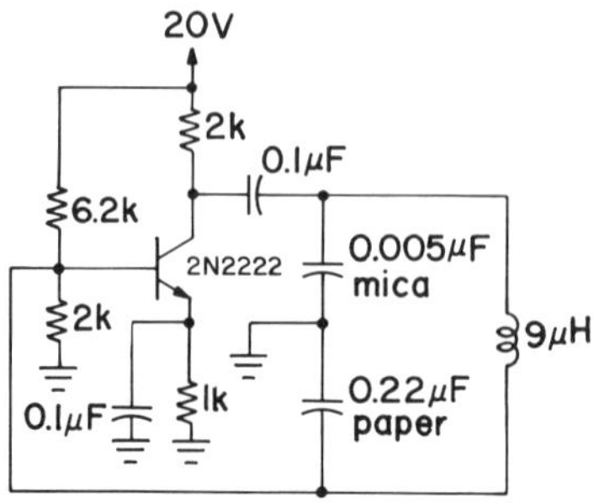

Figure 9.8 A working example of a Colpitts oscillator.

If the reactive elements of the oscillator are chosen to satisfy

$$\frac{L_1 L_2 - M^2}{(L_1 + L_2 + 2M)C} \ll R_i R_o \tag{9.36}$$

then the frequency of oscillation becomes independent of the resistances loading the resonant circuit, and the result produced by (9.22) is identical to that obtained from (9.21). Condition (9.36) is not difficult to obtain in practice, particularly if the coil is very tightly coupled, in which case the numerator of the left-hand side of (9.36) tends to zero.

When the values of (9.33) are substituted into (9.26), we find that the voltage gain must satisfy

$$g_m R_o \geq \frac{L_1 + M}{L_2 + M} \tag{9.37}$$

Example 9.2: A working model of a Colpitts oscillator is shown in Fig. 9.8. This oscillator was built with readily available components. The air core inductor consists of 21 turns of 16 gauge copper wire wound on a 25.4 mm form. Its Q was measured at 66. The 0.22 μF capacitor was selected by trial and error to give the biggest output for the lowest possible distortion. The calculations associated with this oscillator are left to an exercise at the end of this chapter. ■

9.5 *Oscillator Crystal Control*

For very good frequency stability, the inductor in a Colpitts oscillator (or any oscillator requiring an inductance) can be replaced by a quartz crystal. As we shall see presently, a quartz crystal appears inductive for a very narrow band of frequencies only, so the frequency of operation of the oscillator must confine itself to that band. The symbol and the equivalent circuit are shown in Figs. 9.9*a* and 9.9*b*, respectively.

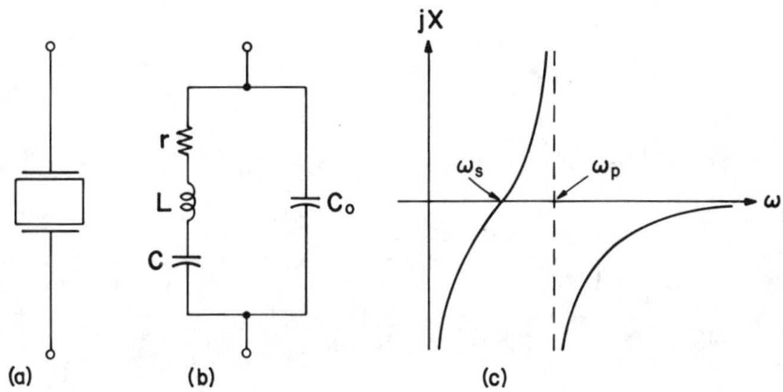

Figure 9.9 (*a*) Quartz-crystal symbol. (*b*) Equivalent circuit. (*c*) Reactance function.

If the resistance r is assumed to have a negligible effect on the operation of the crystal at its operating frequency, then the crystal is seen as a purely reactive element. Its reactance as a function of frequency is shown in Fig. 9.9*c*. The two critical frequencies associated with that plot are

$$\omega_s^2 = \frac{1}{LC} \tag{9.38}$$

and

$$\omega_p^2 = \omega_s^2\left[1 + \frac{C}{C_0}\right] \tag{9.39}$$

The Q of the crystal is the quality factor of the inductance and is given by

$$Q = \frac{\omega_s L}{r} \tag{9.40}$$

and is usually well in excess of 1000.

Example 9.3: Some typical values for a clock oscillator crystal are: f_s = 32.768 kHz, C_0 = 1.7 pF, C = 0.0034 pF, and r = 30 kΩ.

From (9.38) we find that L = 6940 H, and from (9.40) we calculate Q = 47617. From (9.39) it follows that $f_p = 1.001 f_s$, hence the crystal appears inductive within a bandwidth of only 32.75 Hz, which lies to the right of f_s. ■

9.6 *The RC Phase-Shift Oscillator*

The Colpitts and Hartley oscillators are commonly used at frequencies above the audio band. As frequency is decreased, then it is readily apparent from (9.13) that the inductor needed for one of these oscillators becomes

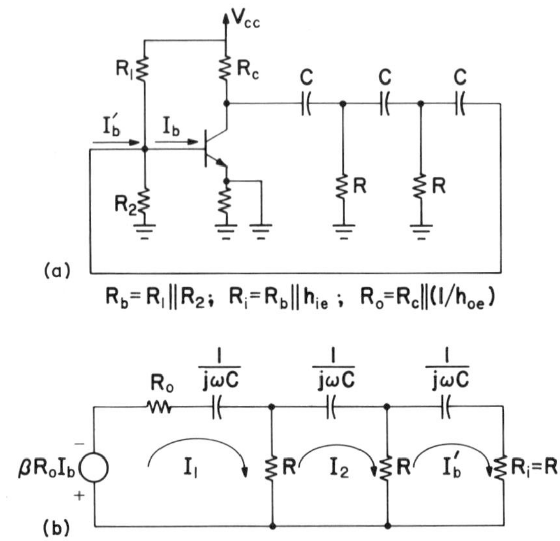

Figure 9.10 The *RC* phase-shift oscillator.

impractically large if a high value of Q is to be obtained for the resonant circuit in the presence of a fixed series resistance. At audio frequencies the oscillators usually employ *RC* phase-shift networks.

In Fig. 9.10*a* we have a bipolar-transistor version of the *RC* phase-shift oscillator. The equivalent circuit for the derivation of the conditions of oscillation is shown in Fig. 9.10*b*. The source on the left-hand side of the equivalent circuit has been changed to a Norton equivalent in anticipation of using mesh analysis to arrive at the expression for the circuit loop gain. After some algebra (which has been left for an exercise), the expression for the oscillator return ratio can be shown to have the form

$$T = \frac{\beta R_o R^2}{3R^2R_o + R^3 - \dfrac{R_o + 5R}{\omega^2C^2} + \dfrac{1}{j\omega C}\left[4RR_o + 6R^2 - \dfrac{1}{\omega^2C^2}\right]} \frac{R_b}{R_b + h_{ie}} \tag{9.41}$$

Setting the imaginary part of the last equation to zero produces the result

$$\omega_0 = \frac{1}{RC\sqrt{6 + 4(R_o/R)}} \tag{9.42}$$

When this is substituted back into (9.41) and condition (9.1a) is applied, we obtain

$$\beta \geq \left[29\frac{R}{R_o} + 4\frac{R_o}{R} + 23\right]\left[1 + \frac{h_{ie}}{R_b}\right] \tag{9.43}$$

The right-hand side of the last equation can be minimized by choosing

$$\frac{R_o}{R} = \frac{\sqrt{29}}{2} \approx 2.69 \tag{9.44}$$

in which case (9.42) and (9.43) reduce to

$$\omega_0 \approx \frac{1}{4.09RC} \tag{9.45}$$

and

$$\beta \geq 44.54\left[1 + \frac{h_{ie}}{R_b}\right] \tag{9.46}$$

The minimum obtained by choosing R_o/R according to (9.44) is very flat, and it can be readily shown that departures from the ideal of as much as 30% have a very small effect on (9.43) but a somewhat larger effect on (9.42). The optimum choice of the resistor ratio for FET oscillators of this type is different from the above result. The analysis of the FET phase-shift oscillator is left as an exercise.

The advantage of the *RC* phase-shift oscillator is that it is very easy to construct using a single low-gain transistor. The disadvantage is that if it is desired to change the operating frequency, three capacitors must be varied. As with all *RC* oscillators, there is no frequency-selective network at the collector to remove distortion components of the collector signal, hence it is very difficult to obtain a clean sinusoidal output using this kind of oscillator.

9.7 *The Twin-T RC Phase-Shift Oscillator*

The twin-*T* *RC* phase-shift oscillator shown in Fig. 9.11 is another configuration that can be used at audio frequencies. It is best to have a high gain amplifier as the active element in order for the relationships which are derived to be accurate. It is assumed that the left-hand side of the twin-*T* network is driven by a very low impedance voltage source, and that the right-hand side sees a very high amplifier input impedance, so that loading on the network is negligible. The resistors R_2 and R_3 are included to obtain

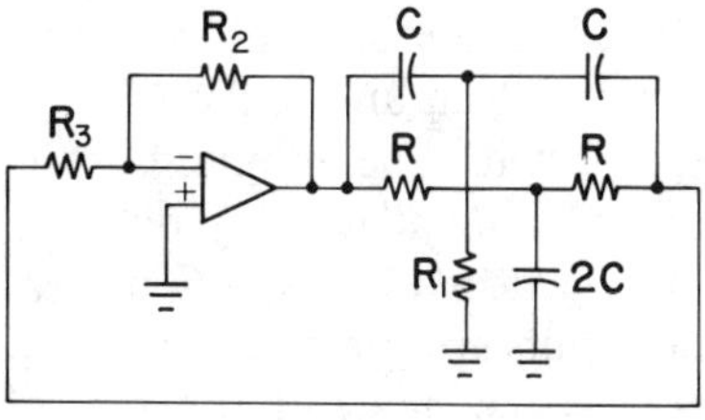

Figure 9.11 The twin-*T* *RC* phase-shift oscillator.

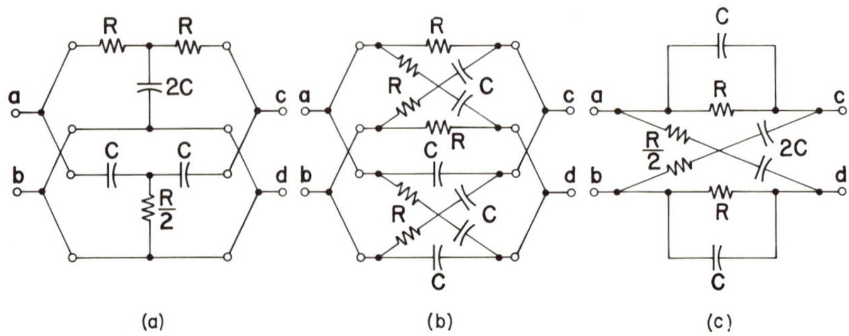

Figure 9.12 Twin-*T* to lattice transformation.

some control over the amplifier gain. The analysis proceeds most easily from Fig. 9.12.

It is very difficult to find the ratio of open-circuit output voltage to input voltage in the twin-*T* network as it appears in Fig. 9.12*a*. But a symmetrical-*T* network can be changed into a balanced lattice using a well-known network transformation [1]. Figure 9.12*b* shows each of the parallel-*T* equivalent networks of Fig. 9.12*a* transformed into a lattice. Since the two lattice networks are in parallel, the two can be combined into one, as shown in Fig. 9.12*c*. If a voltage source is connected at terminals *a–b*, then the voltage at terminals *c–b* is obtained by using voltage division. Similarly voltage division is used for finding the voltage at terminals *d–b*. The voltage at *c–d* is the difference of the two previously obtained voltages. After some simplification, it is found that the voltage transfer function for this network is

$$H(j\omega) = \frac{1 - (\omega/\omega_0)^2}{1 - (\omega/\omega_0)^2 + 4j(\omega/\omega_0)} \tag{9.47}$$

where

$$\omega_0 = 1/RC \tag{9.48}$$

It is apparent from the above equation that this network provides a $-90°$-phase shift for input signals whose frequency is slightly less than ω_0 and a $-270°$-phase shift for input signals whose frequency is slightly greater than ω_0. At ω_0 the magnitude of $H(j\omega)$ equals zero and it can be reasonably argued that the phase shift is $-180°$. Such an oscillator could not work since the amplifier gain cannot be expected to be infinite. For the oscillator of Fig. 9.11 to work, the twin-*T* must be unbalanced by reducing R_1 below the ideal value of $R/2$. The consequences of this action were analyzed by using a digital computer and the results are summarized in Fig. 9.13. For realistic choices of the resistor R_1, the actual frequency of

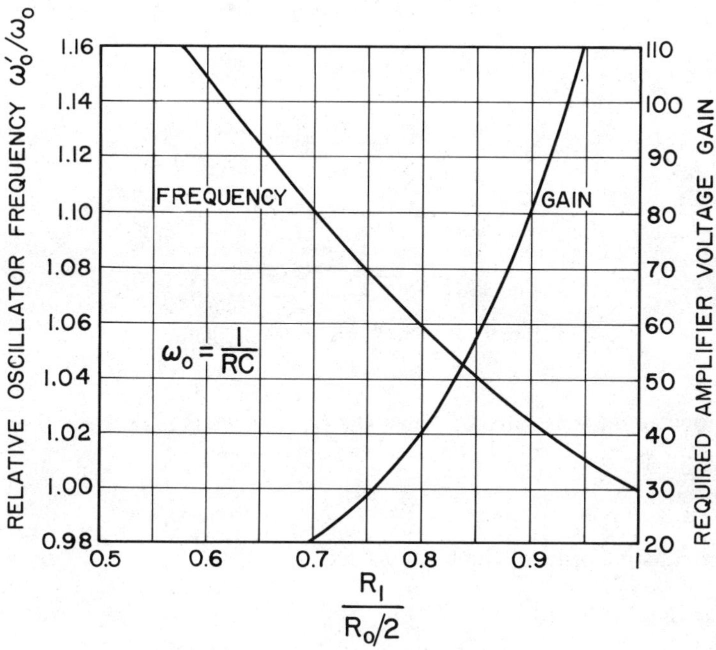

Figure 9.13 The consequences of reducing R_1 below $R/2$.

operation, designated ω_0', is increased relative to the ideal frequency ω_0. The amplifier gain required to obtain oscillation is also shown. In practice a high gain amplifier is used and R_1 is made slightly adjustable.

As with the phase-shift oscillator, the disadvantage with the twin-*T* oscillator is that if it is desired to change the operating frequency, three capacitors must be varied. The advantage is that the closer R_1 is to its ideal value of $R/2$, the greater is the phase variation of $H(j\omega)$ with respect to ω. So this oscillator can be made very frequency stable if the components of the twin-*T* network are stable.

9.8 *The Wien-Bridge Oscillator*

This oscillator consists of an amplifier whose output is connected to the input by means of a Wien bridge as shown in Fig. 9.14. The FET labelled R_2 is used to provide the resistance needed for the fourth arm of the Wien bridge, and its presence will be explained later.

For the conditions of oscillation to be met, the bridge is operated near balance. This occurs at the frequency at which

$$R_2\left[R + \frac{1}{j\omega C}\right] = R_1\left[\frac{R}{1 + j\omega CR}\right] \qquad (9.49)$$

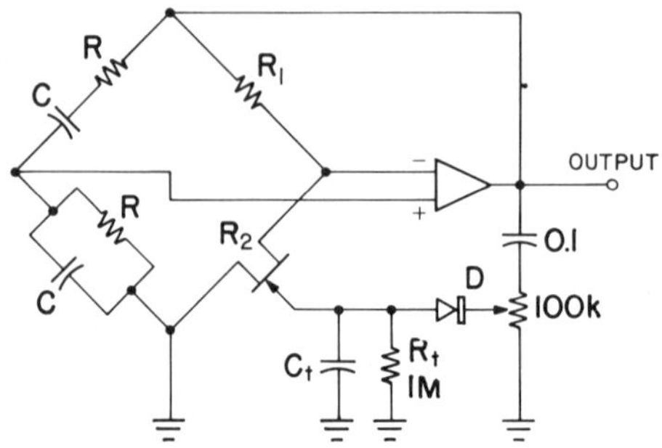

Figure 9.14 The Wien-bridge oscillator.

The above equality is satisfied when the frequency takes on the value

$$\omega_0 = 1/RC \tag{9.50}$$

and the resistive arms of the bridge satisfy

$$R_1 = 2R_2 \tag{9.51}$$

When the bridge is perfectly balanced, there is no input to the amplifier, so that in practice R_2 is made slightly smaller than the value determined from (9.51), in order to obtain some amplifier drive. If the bridge is unbalanced too much, then the output becomes highly distorted, but if the unbalance is insufficient, then the oscillations cease. Some circuitry is usually added to the oscillator to assure that the bridge is maintained very near the balance point. In this case, the circuitry at the oscillator output, which appears in Fig. 9.14, applies a fraction of the negative of the peak output voltage to the gate of the FET labelled R_2. If the amplifier output becomes too great, then the FET is driven closer to pinchoff, raising its resistance, thus bringing the bridge closer to balance, causing a reduction in the oscillator output. The converse is true if the amplitude of the oscillations decreases. Numerous other schemes exist for stabilizing the oscillator amplitude output.

The advantage of this oscillator is that only two capacitors have to be changed if a frequency range change is desired. The frequency within the range can be controlled by varying the values of the two bridge resistors labelled R. The Hewlett–Packard company was founded on a single product—the Wien-bridge vacuum-tube audio oscillator. A problem that had to be solved was the attainment of good amplitude control with simple circuitry. In that product, a light bulb (which is a positive coefficient temperature-dependent resistor) was used in place of R_2 to get the desired level of amplitude control.

9.9 *Frequency Stabilization Factor*

When an oscillator is in operation, the angle of the loop gain, which we designate ϕ_T, is equal to $-180°$. Let us suppose, for the sake of argument, that the phase angle of the amplifier changes by some amount $-\Delta\phi_T$. The phase-shift network has to compensate for this change in amplifier angle by an amount $+\Delta\phi_T$ by changing the frequency of oscillation by a relative shift in operating frequency of value $\Delta\omega/\omega$. We see that it is sensible to define a frequency stabilization factor of the form

$$S_f = \left.\frac{\Delta\phi_T}{\Delta\omega/\omega}\right|_{\omega=\omega_0} = \omega_0 \left.\frac{\Delta\phi_T}{\Delta\omega}\right|_{\omega=\omega_0} \tag{9.52}$$

If S_f is a large number, then that implies that a small value of $\Delta\omega/\omega$ can be used to compensate for a large change in $\Delta\phi_T$. In differential form the above becomes

$$S_f = \omega_0 \left.\frac{d\phi_T}{d\omega}\right|_{\omega=\omega_0} \tag{9.53}$$

Example 9.4: In an FET Colpitts oscillator we assume that $R_i \to \infty$, hence (9.20) reduces to

$$T(j\omega) = \frac{g_m R_o Z_1 Z_2}{Z_1(Z_2 + Z_3) + R_o(Z_1 + Z_2 + Z_3)} \tag{9.54}$$

After substituting the values in (9.28) in the above, we have

$$T(j\omega) = \frac{g_m R_o}{1 - \omega^2 L C_2 + j\omega(C_1 + C_2 - \omega^2 L C_1 C_2) R_o} \tag{9.55}$$

To meet the condition of oscillation, the term within the parentheses in the denominator must vanish at ω_0 namely

$$(C_1 + C_2 - \omega_0^2 L C_1 C_2) = 0 \tag{9.56}$$

from which it follows that the frequency of oscillation is given by

$$\omega_0^2 = \frac{C_1 + C_2}{L C_1 C_2} \tag{9.57}$$

The angle of $T(j\omega)$ in (9.55) is given by

$$\tan\phi_T = -\frac{\omega(C_1 + C_2 - \omega^2 L C_1 C_2) R_o}{1 - \omega^2 L C_2} \tag{9.58}$$

which, at the frequency of oscillation, evaluates to

$$\phi_T = -180^\circ \qquad \text{when} \qquad \omega = \omega_0 \tag{9.59}$$

Differentiating both sides of (9.58) with respect to ω, and using (9.59), (9.56), and (9.57), it easily follows that

$$\left.\frac{d\phi_T}{d\omega}\right|_{\omega=\omega_0} = \frac{2\omega_0^2 LC_1C_2}{1 - \omega_0^2 LC_2} R_o \tag{9.60}$$

Substituting into (9.53) and using (9.57) we get

$$S_f = -2\omega_0 C_1 \left[\frac{C_1 + C_2}{C_2}\right] R_o \tag{9.61}$$

From (9.17) we get for this case

$$R_{eq} = \left[\frac{C_1 + C_2}{C_2}\right]^2 R_o \tag{9.62}$$

and for the resonant circuit used in this case, the equivalent capacitance is the series combination of C_1 and C_2, namely

$$C_{eq} = \frac{C_1C_2}{C_1 + C_2} \tag{9.63}$$

When the last two equations are substituted into (9.61), we obtain

$$S_f = -2\omega_0 C_{eq} R_{eq} \tag{9.64}$$

From (9.4) we recognize that the above can be written in the form

$$S_f = -2Q \tag{9.65}$$

The higher the quality factor of the resonant circuit the higher is the frequency-stabilization factor. It therefore comes as no surprise that high-quality Colpitts oscillators are usually built with quartz crystals, since their Q is orders of magnitude greater than that of magnetic inductors. ■

REFERENCE

1. Guillemin, E. A.: "Synthesis of Passive Networks," John Wiley, New York, 1957, p. 203.

EXERCISES

9.1. An inductor L is available whose quality factor is Q_1. The inductor is shunted by a resistor of value R_2 and we define $Q_2 = R_2/\omega_0 L$.

(a) Calculate the equivalent resistance shunting the coil and show that the Q of the entire circuit is given by

$$\frac{1}{Q} = \frac{1}{Q_1} + \frac{1}{Q_2}$$

(b) Can this result be generalized to any number of resistors shunting the coil?

9.2. In the working example of a Colpitts oscillator shown in Fig. 9.8, the transistor is biased for a collector current of 4 mA and its current gain is 100.

(a) Find the resonant frequency of the oscillator.

(b) Find the output resistance R_o and the input resistance R_i for the circuit. Lump the effects of the inductor loss into R_o.

(c) Find the Q of the resonant circuit, accounting for all resistive loading.

(d) By how great a margin is (9.32) satisfied?

(e) Redesign this oscillator with the idea that the second term on the right-hand side of (9.24) be used to balance the left-hand side of the equation by the same margin as was found in part (d). Find the new values of C_1 and C_2.

(f) What is the overall Q of the resonant circuit for this design?

9.3. A typical crystal has the parameters $L = 1.6$ H, $C_0 = 10$ pF, $C_0/C = 355$, and $Q = 10^6$.

(a) Find f_s, f_p, and r for this crystal and draw the equivalent circuit.

(b) Ignoring r, write an expression for the driving point impedance of the crystal.

(c) At what frequency is the impedance of part (c) equivalent to a 9-μH inductance (which could be used in the oscillator of Example 9.2)?

9.4. Using the equivalent circuit for the transistor phase-shift oscillator shown in Fig. 9.10*b* derive an expression for the return ratio of the *RC* phase-shift oscillator.

9.5. (a) Verify the connection between (9.43) and (9.44) and (9.45) and (9.46).
(b) Computationally show that the claim in the paragraph following (9.46) is correct.

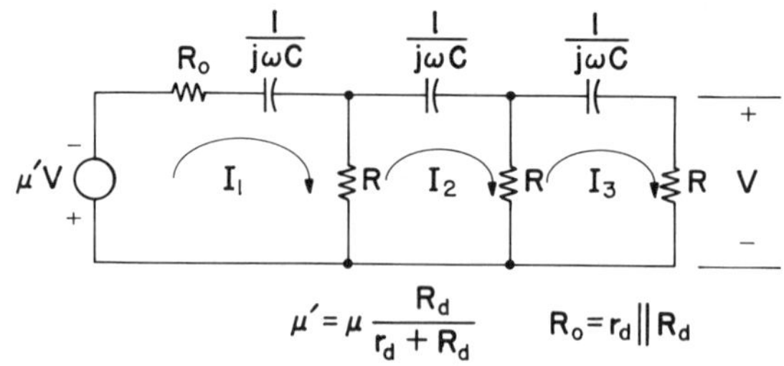

Figure P9.6

9.6. The equivalent circuit for an FET phase-shift oscillator is shown in Fig. P9.6.
(a) Derive the expression for return ratio.
(b) Find the conditions of oscillation.
(c) What is the best choice of R relative R_o so that the oscillator will work with FETs with low values of μ? What is the corresponding value of ω_0?

9.7. In Fig. 9.12*a* the resistor labelled $R/2$ in the lower T network is replaced by a resistor whose value is $\alpha R/2$.
(a) Find the lattice equivalent of this new twin-T network.
(b) Write an expression for the voltage transfer function $H(j\omega)$.
(c) Verify some of the points in Fig. 9.13.

Hint: It is advisable to use a computer or a programmable calculator for this purpose.

9.8. In the Wien-bridge oscillator of Fig. 9.14, assume the amplifier has an infinite input impedance, zero output impedance, and a voltage gain A. Assume the FET is replaced by a resistor of value $aR_1/2$.
(a) Find the expression for the return ratio $T(j\omega)$.
(b) Apply the conditions of oscillation to find the oscillator operating frequency and the required amplifier gain.
(c) Derive the expression for S_f (see section 9.9).

A

Transistor Equivalent Circuit Transformations

IT IS SOMETIMES DIFFICULT TO CALCULATE the gain or impedances for a bipolar-transistor or FET connection, without getting involved in tedious simultaneous equations solutions. This is usually a consequence of the need to use mesh or nodal equations for solving these problems, methods which need not be resorted to when the proper equivalent circuit transformations are used. Four equivalent circuit transformations will be presented in this appendix. Two will apply to bipolar transistors and two will apply to FETs. Their use will facilitate the analysis of feedback amplifiers.

Bipolar-Transistor Equivalent Circuit

The schematic symbol and the small signal equivalent circuit for a bipolar transistor for the low-frequency case are shown in Fig. A.1. The currents and voltages shown denote small (incremental) signals for a transistor.

In practice, it is found that the parameters h_{oe} and h_{re} do not significantly affect the final answers when performing transistor circuit analysis [1], and so the simpler model of Fig. A.2 can be used for greater convenience. Throughout this book the common emitter model will be used for bipolar transistors, so that the "e" subscript will be suppressed in all bipolar-transistor diagrams which follow.

The parameter h_f, sometimes referred to as β, links I_c to I_b by the equation

$$I_c = h_f I_b = \beta I_b \tag{A.1}$$

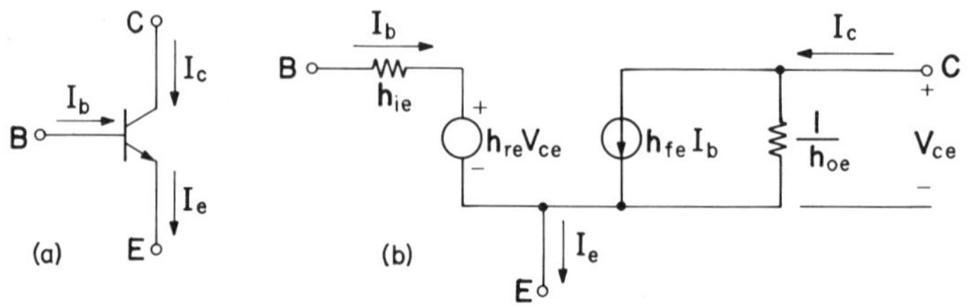

Figure A.1 Schematic representation (*a*) and small signal equivalent circuit (*b*) for a bipolar transistor.

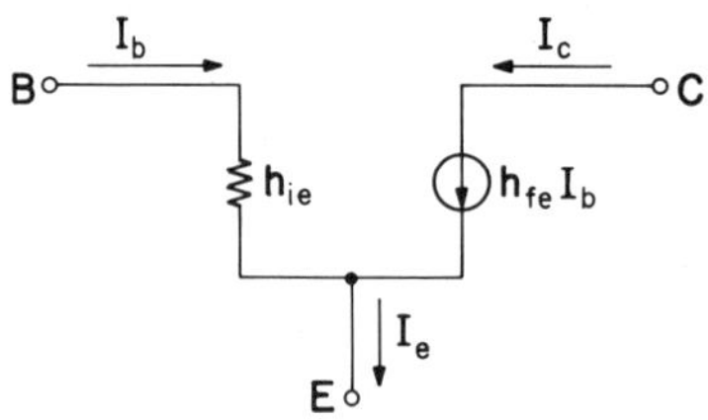

Figure A.2 Simplified equivalent circuit for bipolar transistors.

From Fig. A.2 it is clear that

$$I_e = I_b + I_c \tag{A.2}$$

Substituting (A.1) into (A.2) we get,

$$I_e = (1 + h_f)I_b = (1 + \beta)I_b \tag{A.3}$$

We can furthermore establish that

$$I_c = \alpha I_e \tag{A.4}$$

where we define

$$\alpha = \frac{h_f}{1 + h_f} = \frac{\beta}{1 + \beta} \tag{A.5}$$

Since β is substantially bigger than unity, it follows that α is very close to unity.

When utilizing small signal equivalent circuits to solve small signal problems, direct voltages and direct currents are set to zero when drawing equivalent circuits.

Example A.1: We wish to find the gain $G = V_o/V_i$ for the two-stage amplifier shown in Fig. A.3.

To solve this problem we draw the small signal model for the two-transistor amplifier as shown in Fig. A.4.

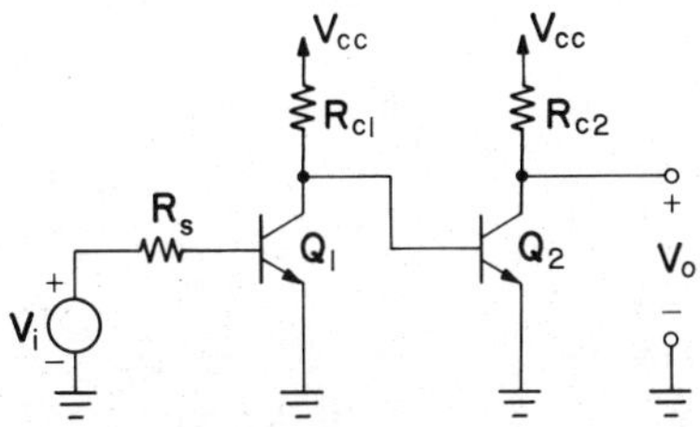

Figure A.3 Two-stage transistor amplifier.

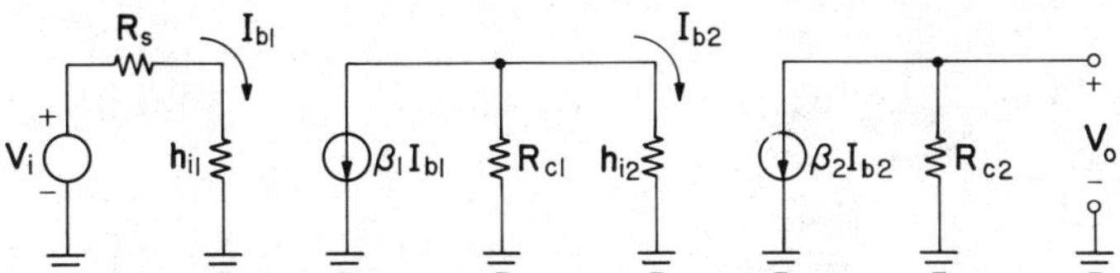

Figure A.4 Equivalent circuit of the two-stage amplifier.

Using current division, we can obtain directly from Fig. A.4,

$$G = \beta_1 \beta_2 \frac{R_{c2}}{R_s + h_{i1}} \frac{R_{c1}}{R_{c1} + h_{i2}} \tag{A.6}$$

■

Equivalent Circuit Transformations

When dealing with complicated circuits, the basic equivalent circuit discussed in the previous section can lead to a great deal of tedious work. We shall take the transistor equivalent circuit a step further in an effort to achieve some simplification. In Fig. A.5 we have a transistor configuration with a very general circuit appearing below the emitter. We are interested in finding its equivalent circuit from the point of view of an observer looking at the base side of the transistor at terminals B–G.

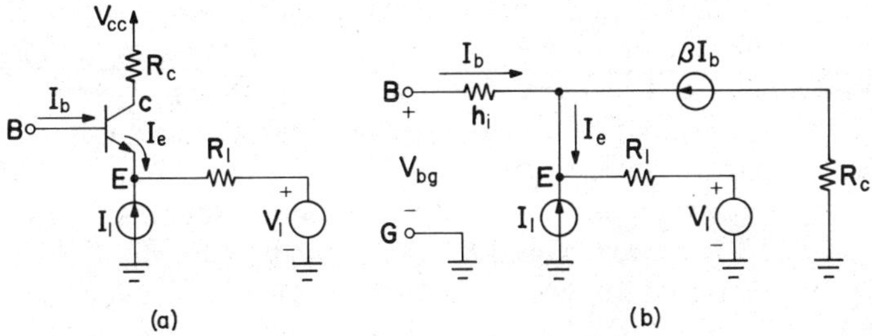

Figure A.5 Diagrams used for deriving equivalent circuits with respect to the transistor base.

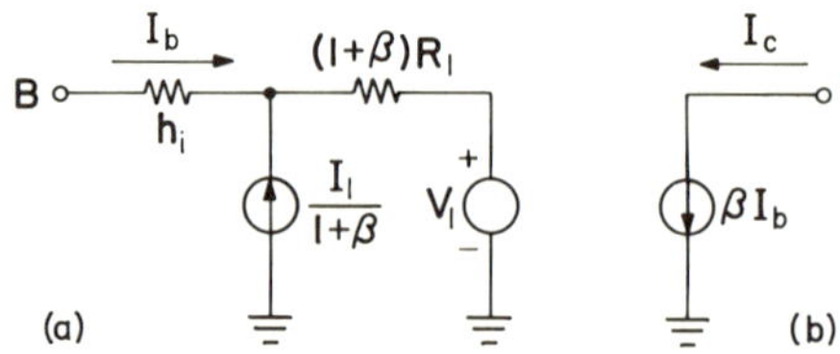

Figure A.6 Equivalent circuit of a bipolar transistor. (*a*) From the point of view of the transistor base. (*b*) As seen at the collector.

When terminals B–G are left open circuited, we observe that I_b is zero, hence I_e and I_c are also zero. Using superposition we readily obtain the open circuit voltage at terminals B–G:

$$\left(V_{bg}\right)_{\text{oc}} = V_1 + R_1 I_1 \tag{A.7}$$

When node B is shorted to ground, we solve for I_b using superposition,

$$(I_b)_{\text{sc}} = -\frac{R_1}{R_1 + h_i}\beta(I_b)_{\text{sc}} - \frac{R_1}{R_1 + h_i}I_1 - \frac{1}{R_1 + h_i}V_1 \tag{A.8}$$

Solving the above for $(I_b)_{\text{sc}}$, we obtain

$$(I_b)_{\text{sc}} = -\frac{V_1}{h_i + (1 + \beta)R_1} - \frac{I_1}{1 + \beta}\frac{(1 + \beta)R_1}{h_i + (1 + \beta)R_1} \tag{A.9}$$

The last equation and (A.7) characterize the circuit shown in Fig. A.6.

The most noteworthy feature of the equivalent circuit of Fig. A.6*a* is that the collector current source βI_b does not appear in it. It is not needed in this equivalent circuit, which is used to perform calculations on the base side of the transistor, and it would be incorrect to include it in the diagram. To make calculations on circuitry that is connected to the collector, Fig. A.6*b* is totally adequate. It is perfectly acceptable to draw the collector current source connected to ground, since the impedance seen at the far side of the current source (at the emitter) does not affect the calculations of voltage or current at the collector.

The equivalent circuit of Fig. A.6 has other noteworthy features. It raises the levels of all impedances connected below the emitter by a factor of $(1 + \beta)$ and reduces currents by the same factor. Voltages are unchanged by this transformation. The current I_e cannot be shown in this diagram, however, once I_b is calculated then I_e can be found using (A.3).

We can arrive at the same kind of equivalent circuit transformation for the transistor from the point of view of an observer standing at the emitter of the transistor. The circuits used for this derivation are shown in Fig. A.7.

The procedure used follows along lines similar to that used in obtaining (A.7) and (A.9). We again use superposition to obtain

$$\left(V_{eg}\right)_{\text{oc}} = V_2 + I_2 R_2 \tag{A.10}$$

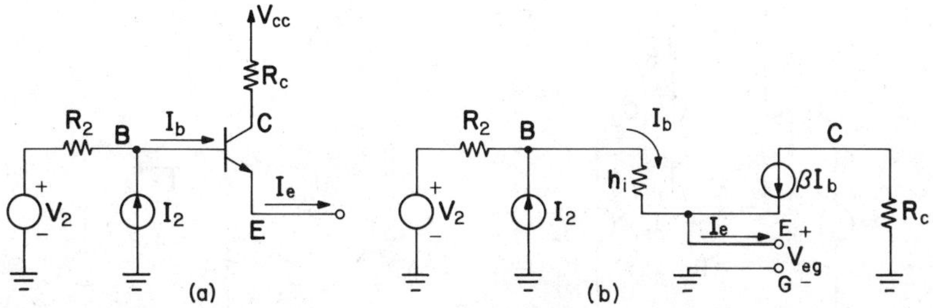

Figure A.7 Diagrams used for deriving equivalent circuits with respect to the transistor emitter.

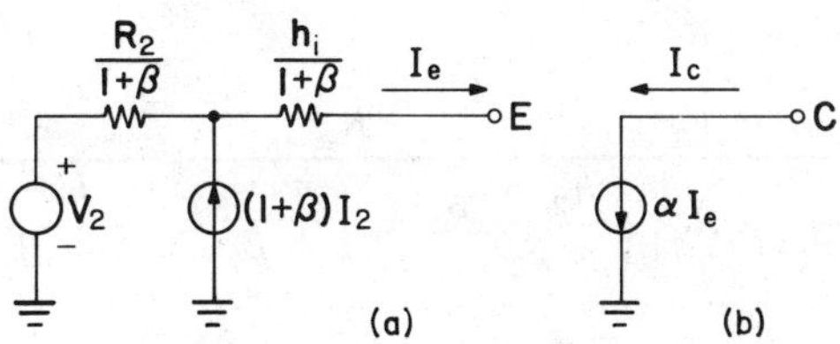

Figure A.8 Equivalent circuit of a bipolar transistor. (*a*) From the point of view of the transistor emitter. (*b*) As seen at the collector.

and

$$(I_e)_{sc} = (1+\beta)\frac{V_2}{R_2 + h_i} + (1+\beta)\frac{R_2}{R_2 + h_i} I_2 \qquad \text{(A.11)}$$

The last two equations characterize the equivalent circuit shown in Fig. A.8. We note that the collector current source does not appear in Fig. A.8*a*. It is not needed in the equivalent circuit, which is used for performing calculations on circuitry connected to the emitter of the transistor, and it would be incorrect to show it. To make calculations on circuitry connected to the collector we can make use of the circuit shown in Fig. A.8*b*.

In this equivalent circuit, all impedances connected above the emitter are lowered by $(1 + \beta)$, whereas all currents above the emitter are increased by the same factor. Voltages remain unchanged by this transformation. The current I_b cannot be shown in this diagram, but once I_e is known, I_b can be calculated using (A.3).

Example A.2: Suppose we wish to find the relationship between V_1, V_2, and V_o for the differential amplifier shown in Fig. A.9. The simplest way to solve this problem is to draw the equivalent circuit of Fig. A.9*b*, which is correct for an observer standing at the common point E. Using

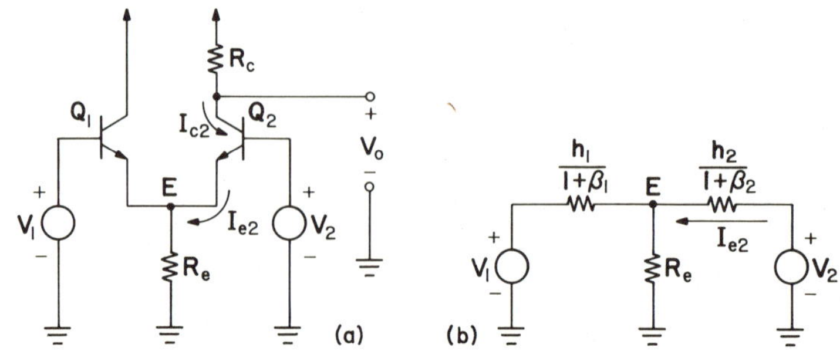

Figure A.9 (*a*) Differential amplifier. (*b*) Equivalent circuit.

superposition we find

$$I_{e2} = \frac{-V_1}{\dfrac{h_1}{1+\beta_1} + R_e \left\| \dfrac{h_2}{1+\beta_2}\right.} \; \frac{R_e}{R_e + \dfrac{h_2}{1+\beta_2}} + \frac{V_2}{\dfrac{h_2}{1+\beta_2} + R_e \left\| \dfrac{h_1}{1+\beta_1}\right.} \tag{A.12}$$

Once the above is calculated, we need only

$$I_{c2} = \alpha_2 I_{e2} \tag{A.13}$$

and

$$V_o = -I_{c2} R_c \tag{A.14}$$

to find the relationship between V_1, V_2, and V_o. This example shows how fairly complicated circuits can be analyzed with ease when the proper transistor equivalent circuits are used. ■

FET Equivalent Circuit Transformations

The schematic symbol and the small signal equivalent circuit for an FET are shown in Fig. A.10. The currents and voltages shown in the diagram denote small (incremental) signals. As was previously done for bipolar transistors, an equivalent circuit will be derived for the FET, which will facilitate analysis of very complicated configurations.

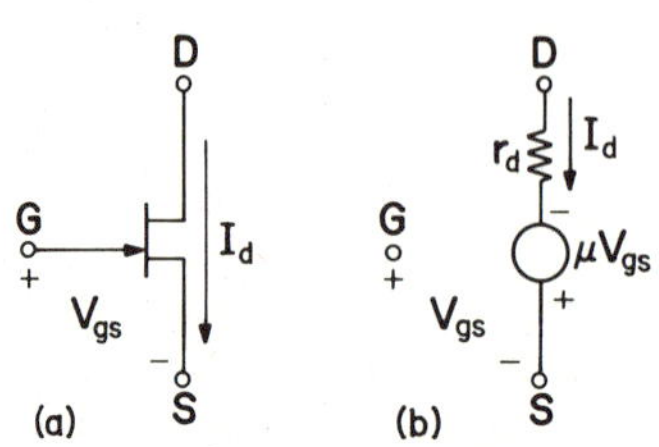

Figure A.10 Schematic representation (*a*) and small signal equivalent circuit (*b*) for an FET.

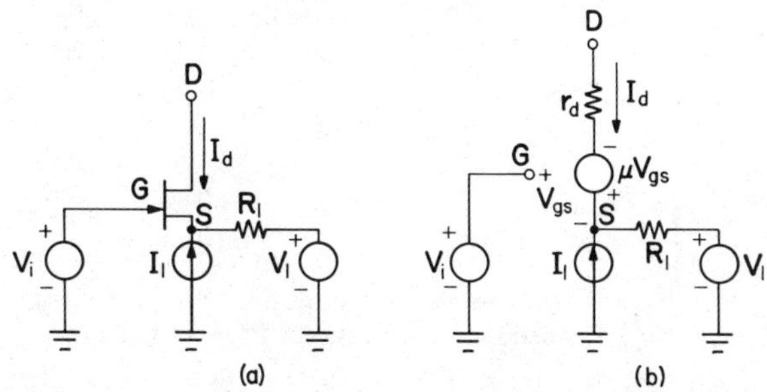

Figure A.11 Schematic diagram (*a*) and equivalent circuit (*b*), for deriving the equivalent circuit with respect to the FET drain terminal.

In Fig. A.11 we have an FET configuration with a general circuit appearing below the source. We are interested in finding an equivalent circuit with respect to the drain terminal *D*. When the terminal *D* is left open circuited, I_d is zero and the voltage at the source terminal *S* is

$$V_s = V_1 + I_1 R_1 \tag{A.15}$$

Hence

$$V_{gs} = V_i - V_s = V_i - V_1 - I_1 R_1 \tag{A.16}$$

The voltage at terminal *D*, which is open circuited, is

$$(V_d)_{\text{oc}} = V_s - \mu V_{gs} \tag{A.17}$$

Substitution of (A.15) and (A.16) into (A.17) yields

$$(V_d)_{\text{oc}} = -\mu V_i + (1 + \mu)(V_1 + I_1 R_1) \tag{A.18}$$

To find the output impedance with respect to terminal *D*, we set all independent sources in Fig. A.11*b* to zero. We attach an external current source of value *I* from *D* to ground, so that the current I_d is equal to *I*, and proceed to calculate the voltage at terminal *D*. The resultant circuit is shown in Fig. A.12.

First we find

$$V_s = IR_1 \tag{A.19}$$

and

$$V_{gs} = -V_s = -IR_1 \tag{A.20}$$

The voltage at terminal *D* is

$$V_d = r_d I - \mu V_{gs} + V_s \tag{A.21}$$

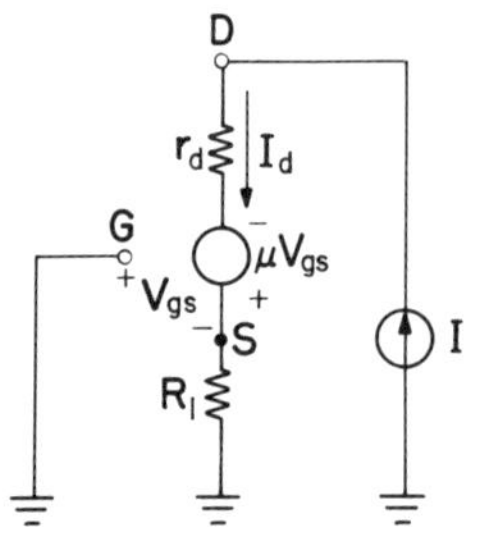

Figure A.12 Resultant circuit for finding the output impedance with respect to terminal *D*.

Substitution of (A.19) and (A.20) into (A.21) yields

$$V_d = \left[r_d + (1 + \mu) R_1 \right] I \tag{A.22}$$

The output impedance at terminal *D* is given by the bracketed portion of (A.22). This output impedance and the open-circuit voltage expression of (A.18) characterize the circuit shown in Fig. A.13.

Substantial simplification is obtained when the equivalent circuit of Fig. A.13 is used to replace the more conventional equivalent circuits of Fig. A.11. A few facts about Fig. A.13 are worth noting. The dependent voltage source is not expressed in terms of the less convenient V_{gs}, but directly in terms of the independent source V_i, which is the source connected from the gate of the FET to ground. All voltage sources and impedances, which are connected below the FET source terminal *S*, are multiplied by a factor $(1 + \mu)$. Currents are unchanged by this transformation. The terminal *S* cannot be shown in Fig. A.13 and this figure cannot be used directly to calculate any voltages below the source terminal *S*. Once I_d is known, however, then this information can be used to proceed with calculations of any desired parameters by reverting to Fig. A.11.

We shall examine the FET equivalent circuit as viewed from the source terminal *S*, as seen in Fig. A.14. We follow a procedure similar to that used in arriving at the results (A.15)–(A.22). In this case we find the expression for $(I_d)_{sc}$ and the output impedance from *S* to ground. Inspection of the equations then shows they characterize the circuit of Fig. A.15.

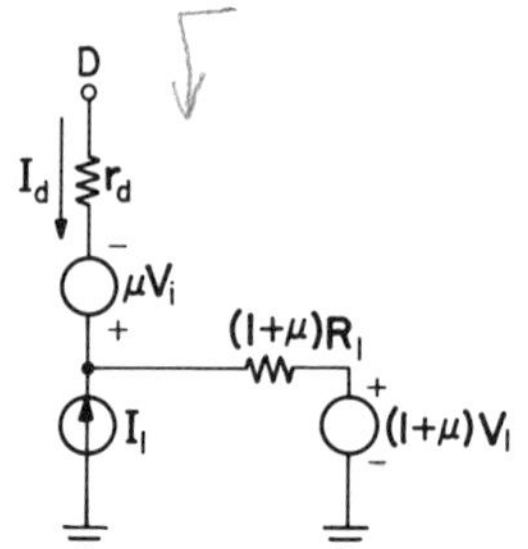

Figure A.13 Equivalent circuit of Fig. A.11 as seen from the drain terminal *D*.

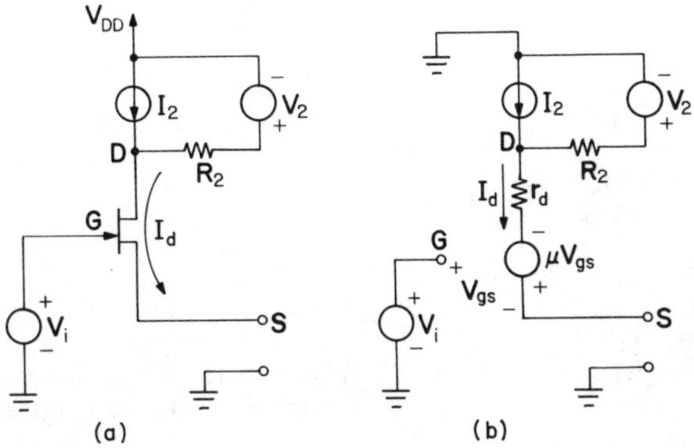

Figure A.14 Diagrams used for deriving the equivalent circuit with respect to the FET source terminal.

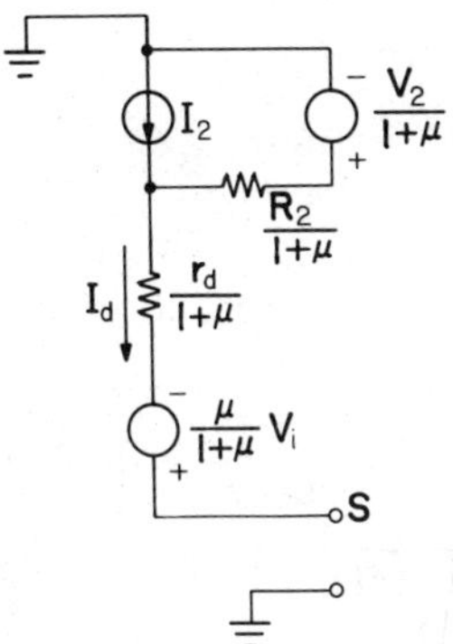

Figure A.15 Equivalent circuit of Fig. A.14 as seen from the source terminal S.

This circuit is much easier to use than the circuit of Fig. A.14. The voltage V_{gs} does not appear in the circuit, hence the solution can be written directly in terms of V_i and any other sources appearing in the circuit.

Example A.3: We wish to find the relationship between V_o and V_1, V_2, and V_3 for the cascode amplifier shown in Fig. A.16.

This problem is easily solved by replacing Q_1 with an equivalent circuit appropriate for an observer looking at its drain terminal, and replacing Q_2 with an equivalent circuit appropriate for an observer looking at its source terminal, as shown in Fig. A.17.

By use of superposition and voltage division, we readily find that

$$V_o = \frac{R_2 + r_{d1} + (1+\mu_1)R_1}{R_{\text{tot}}} \frac{\mu_2}{1+\mu_2} V_2 + \frac{(R_3 + r_{d2})/(1+\mu_2)}{R_{\text{tot}}} [(1+\mu_1)V_3 - \mu_1 V_1] \qquad \text{(A.23)}$$

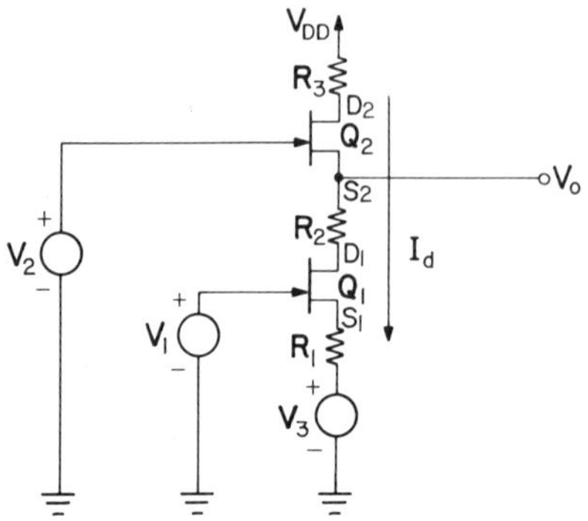

Figure A.16 Cascode amplifier excited by three independent voltage sources.

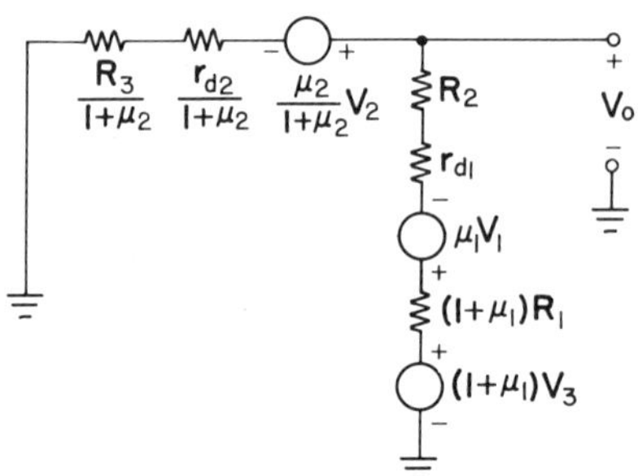

Figure A.17 Cascode amplifier as seen from the output terminal.

where

$$R_{\text{tot}} = \frac{R_3 + r_{d2}}{1 + \mu_2} + R_2 + r_{d1} + (1 + \mu_1)R_1 \tag{A.24}$$

The techniques learned so far can also be used for finding impedances. To find the output impedance for this example, we follow the rule that output impedance is defined as the impedance seen at the output terminals when all independent sources are set to zero. From Fig. A.17 it readily follows that the output impedance Z_o is given by

$$Z_o = [(1 + \mu_1)R_1 + r_{d1} + R_2] \| [(r_{d2} + R_3)/(1 + \mu_2)] \tag{A.25}$$

■

Special Considerations

Many FETs have drain characteristics which are so flat, that r_d can be considered to be infinite in those cases. The techniques described in this section remain applicable for those special cases. The problem is solved as before, but at the conclusion a substitution should be made for μ in terms of r_d and g_m using the well known identity

$$\mu = r_d g_m \tag{A.26}$$

The result is then evaluated in the limit as $r_d \to \infty$. The transconductance

g_m remains finite in the limit, consequently, for that condition, $\mu \to \infty$.

Conclusion

The amount of work that is needed to analyze fairly complicated bipolar-transistor and field-effect transistor structures is greatly reduced when the principles presented in this appendix are applied. When these methods are applied to most feedback amplifier problems, the structures are reduced to the point where finding the solution becomes an almost trivial problem of applying voltage division, current division, and superposition.

REFERENCE

1. Millman J., and Halkias, C. C.: "Electronic Fundamentals and Applications: For Scientists and Engineers," McGraw-Hill, 1976, p. 223.

EXERCISES

A.1. For the differential amplifier of Fig. A.9, find the impedance seen by the input source V_1.

A.2. For the transistors shown in Fig. PA.2, $\beta = 50$ and $h_i = 1k$. Find the gains:
(a) $A_1 = I_{e2}/V_s$.
(b) $A_2 = V_{c1}/V_s$.
(c) $A_3 = V_o/V_s$.

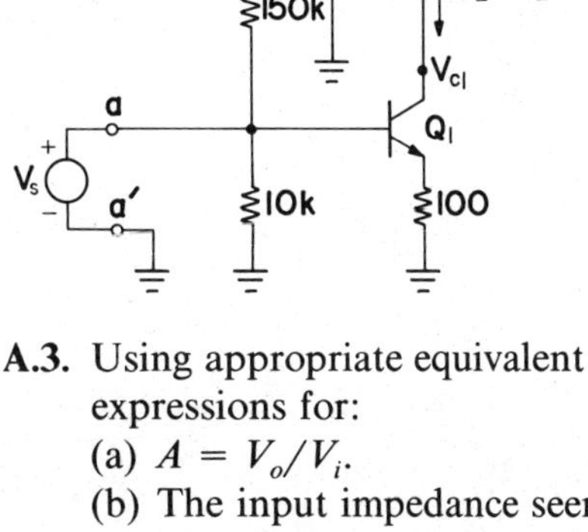

Figure PA.2

A.3. Using appropriate equivalent circuit transformations in Fig. PA.3, find expressions for:
(a) $A = V_o/V_i$.
(b) The input impedance seen by the input source V_i.
(c) The output impedance $Z_{bb'}$.

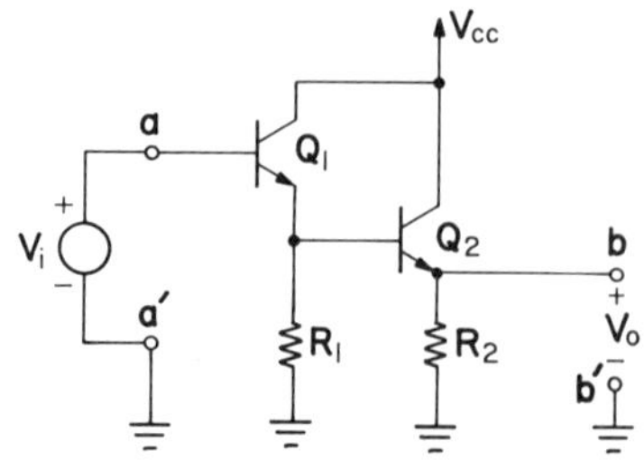

Figure PA.3

A.4. For the transistors shown in Fig. PA.4, $h_i = 1k$ and $\beta = 100$.
(a) Find the gain $A_1 = V_{01}/V_i$.
(b) Find the gain $A_2 = V_{02}/V_i$.
(c) The impedance seen by input source V_i is $Z_{aa'}$. Find $Z_{aa'}$, $Z_{bb'}$, and $Z_{cc'}$.

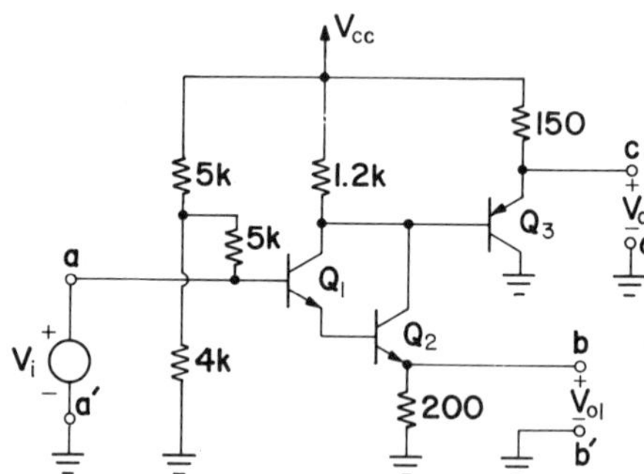

Figure PA.4

A.5. Using appropriate FET equivalent circuit transformations in Fig. PA.5 find:
(a) The voltage gain V_o/V_i.
(b) The input impedance Z_i.
(c) The output impedance Z_o.

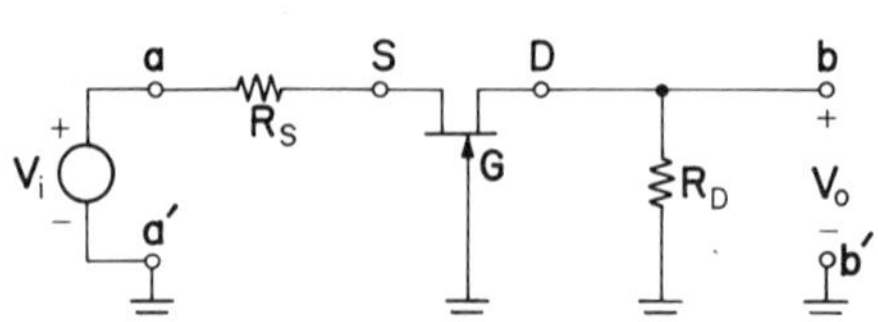

Figure PA.5

A.6. For Fig. PA.6 obtain a simple equivalent circuit suitable for finding V_o, and then find an expression for it.

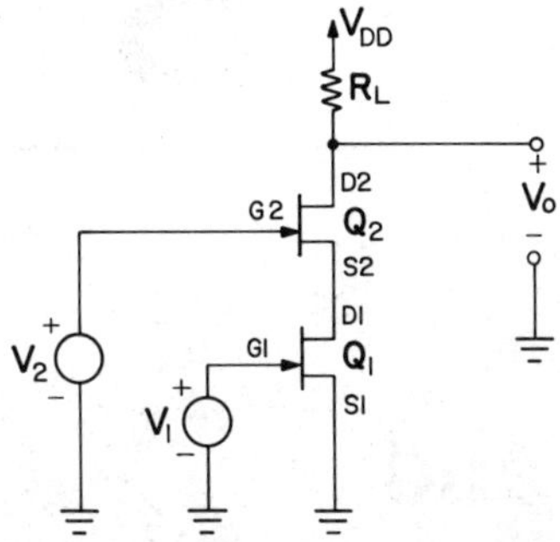

Figure PA.6

A.7. In Fig. PA.7 assume the two FETs are identical. Find:
(a) The gain V_o/V_s.
(b) The output impedance.

Hint: Observe that $G2$ is connected to V_o.

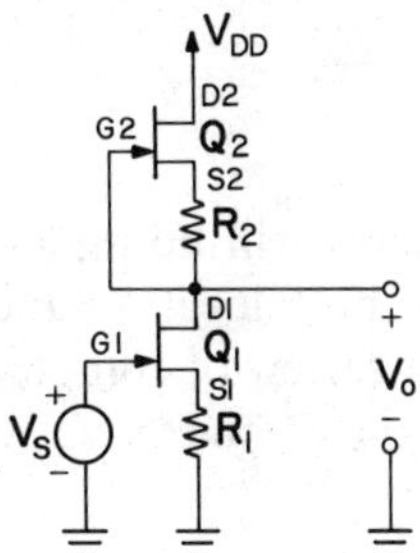

Figure PA.7

A.8. In Fig. PA.8 both FETs have $\mu = 50$ and $r_d = 5k$. V_r represents the power-supply ripple voltage. How much of it will appear at the output?

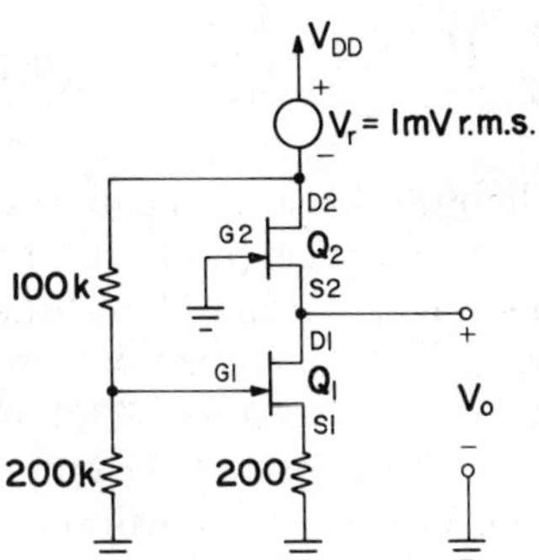

Figure PA.8

A.9. Investigate Eqs. (A.23)–(A.25) as $r_d \to \infty$.

Hint: See the discussion surrounding (A.26)

B

Laplace Transform and Complex Variable Review

IN THIS APPENDIX a short review of the Laplace transform will be undertaken so that the relationship between transfer function pole location and impulse response will be understood. In addition, a short review of complex variable theory will be presented, so that the derivation of the Nyquist criterion in Chap. 6 can be more easily understood.

Laplace Transform and Impulse Response [1, 2]

Consider a linear circuit with the transfer function

$$H(S) = \frac{N(s)}{D(s)} = \frac{b_m s + b_{m-1} s + \cdots + b_1 s + b_0}{a_m s + a_{m-1} s + \cdots + a_1 s + a_0} \tag{B.1}$$

It is well known that networks consisting of lumped, linear elements have transfer functions, which are the ratios of two polynomials (rational functions) of the complex variable s, with real coefficients $\{a_k\}$ and $\{b_k\}$. If the input to the network is a unit impulse $\delta(t)$, which has a Laplace transform of unity, then the output will be the product of the input transform and the transfer function $H(s)$. It is clear that the Laplace transform of the output is $H(s)$ when the input is a unit impulse $\delta(t)$. The inverse transform of $H(s)$, which is $h(t)$, is therefore referred to as the impulse response of the network. The impulse response of the network is determined by the poles of the transfer function $H(s)$. A short review will be presented on how to find the impulse response from a known transfer function.

Transfer Functions with Simplex Poles. The simplest case to consider is that of transfer functions with first order (simple) poles only. Assuming that the poles are located at $p_1, p_2, \ldots, p_n$, then $H(s)$ can be written in the form

$$H(s) = \frac{N(s)}{(s - p_1)(s - p_2)\ldots(s - p_n)} \tag{B.2}$$

To find the impulse response $h(t)$, $H(s)$ is expanded into a partial fraction expansion of the form

$$H(s) = \frac{A_1}{s - p_1} + \frac{A_2}{s - p_2} + \cdots + \frac{A_n}{s - p_n} \tag{B.3}$$

where the coefficients A_i are found using the equation

$$A_i = (s - p_i)H(s)|_{s=p_i} \tag{B.4}$$

From a knowledge of the fact that the Laplace transforms of exponentials give rise to poles, namely

$$\mathscr{L}[e^{pt}] = \frac{1}{s - p} \tag{B.5}$$

we see that the inverse Laplace transform of (B.3) is given by

$$h(t) = A_1 e^{p_1 t} + A_2 e^{p_2 t} + \cdots + A_n e^{p_n t} \tag{B.6}$$

Example B.1: It is desired to find the impulse response of a network whose transfer function is given by

$$H(s) = \frac{9(s^2 + 7s + 10)}{(s + 1)(s + 4)(s + 7)} \tag{B.7}$$

As a first step, it is desired to put (B.7) into the form shown in (B.3). By the repeated application of (B.4), we find that $A_1 = 2$, $A_2 = 2$, and $A_3 = 5$. The transfer function $H(s)$ can therefore be written as an equivalent partial fraction expansion

$$H(s) = \frac{2}{s + 1} + \frac{2}{s + 4} + \frac{5}{s + 7} \tag{B.8}$$

Using (B.5) it is readily apparent that the impulse response, which is the inverse transform of $H(s)$, is given by

$$h(t) = 2e^{-t} + 2e^{-4t} + 5e^{-7t} \tag{B.9}$$

■

In this example, we had poles located on the negative real axis of the s plane, at $s = -1$, $s = -4$, and $s = -7$. These gave rise to decaying exponentials in the impulse response. Had the poles been located on the positive real axis, then $h(t)$ would have contained rising exponentials and would have represented an unstable network, one whose energy output rises indefinitely. A pole at $s = 0$ would give rise to a constant (dc) output, which is considered undesirable from a stability point of view. We are interested in finding the conditions that are necessary for networks to be stable. We see that for networks to be stable, all real poles must lie on the negative real axis.

Transfer Functions with Multiple Poles. If $H(s)$ has a root of multiplicity r at p_j as shown below

$$H(s) = \frac{N(s)}{(s - p_1)(s - p_2)\dots(s - p_j)^r} \tag{B.10}$$

then it can be written in a partial fraction expansion of the form

$$H(s) = \frac{A_1}{s - p_1} + \frac{A_2}{s - p_2} + \cdots + \frac{A_{j1}}{s - p_j} + \frac{A_{j2}}{(s - p_j)^2} + \cdots + \frac{A_{jr}}{(s - p_j)^r} \tag{B.11}$$

The numerator coefficients for the simple poles can be evaluated using (B.4), but for the multiple poles we use

$$A_{jn} = \frac{1}{(r - n)!}\frac{d^{r-n}}{ds^{r-n}}\left[(s - p_j)^r H(s)\right]\Big|_{s=p_j} \tag{B.12}$$

The inverse transform for the simple pole terms is found using (B.5). The inverse transform of the higher-order pole terms is found using

$$\mathscr{L}\left[\frac{t^{n-1}}{(n - 1)!}e^{pt}\right] = \frac{1}{(s - p)^n} \tag{B.13}$$

It is readily apparent that multiple poles produce polynomials multiplied by exponentials.

Example B.2: The application of the above is very simple. Take, as an example,

$$H(s) = \frac{2s^2 + 3s + 7}{(s + 1)^3} = \frac{A_{11}}{s + 1} + \frac{A_{12}}{(s + 1)^2} + \frac{A_{13}}{(s + 1)^3} \tag{B.14}$$

To find the numerator coefficients of the partial fraction expansion, we form the expression

$$(s + 1)^3 H(s) = 2s^2 + 3s + 7 = A_{11}(s + 1)^2 + A_{12}(s + 1) + A_{13} \tag{B.15}$$

The value of A_{13} is obtained from (B.15) by setting $s = -1$. It follows that $A_{13} = 6$. Next (B.15) is differentiated with respect to s, to obtain

$$\frac{d}{ds}\left[(s+1)^3 H(s)\right] = (4s+3) = 2A_{11}(s+1) + A_{12} \qquad \text{(B.16)}$$

When we set $s = -1$, we obtain $A_{12} = -1$. Differentiating (B.16) again with respect to s, we get

$$\frac{d^2}{ds^2}(s+1)^3 H(s) = 4 = 2A_{11} \qquad \text{(B.17)}$$

and it follows from this that $A_{11} = 2$. The partial fraction expansion therefore takes the form

$$H(s) = \frac{2}{s+1} - \frac{1}{(s+1)^2} + \frac{6}{(s+1)^3} \qquad \text{(B.18)}$$

The above is inverted using (B.13), and the result is

$$h(t) = [2 - t + 3t^2]e^{-t} \qquad \text{(B.19)}$$

■

It is clear that multiple poles produce exponentials which are multiplied by polynomials. In this case, the pole lies in the left half of the s plane (at $s = -1$), so we have a decaying exponential, which very rapidly overpowers the polynomial as t increases, and we therefore have an impulse response, which goes to zero as t approaches infinity. A multiple pole at $s = 0$ would give rise to an impulse response consisting entirely of a polynomial, which would increase with time and therefore be representative of an unstable network. We see that the claim that real poles must lie on the negative real axis for stable systems is supported by the last example, and that no poles should be allowed on the $j\omega$ axis.

Transfer Functions with Complex Conjugate Poles. The partial fraction expansion treatment when complex conjugate poles occur in $H(s)$ is the same as it is for real poles, but the resulting impulse response is somewhat different. Since the polynomial coefficients in $H(s)$ are real, then complex poles must occur in conjugate pairs. If p_i is a complex pole, then a portion of $H(s)$ will be the partial response $H_i(s)$ which will have the form

$$H_i(s) = \frac{A_i}{s - p_i} + \frac{A_i^*}{s - p_i^*} \qquad \text{(B.20)}$$

The asterisk above signifies the complex conjugate. The right-hand side of the last equation produces a real $H_i(s)$ because of the simple fact that the sum of a complex number and its conjugate, equals twice the real part of either the number or the conjugate,

$$A + A^* = 2\,\mathrm{Re}(A) = 2\,\mathrm{Re}(A^*) \tag{B.21}$$

The inverse Laplace transform of (B.20) is obtained using (B.5), with the result

$$h_i(t) = A_i e^{p_i t} + A_i^* e^{p_i^* t} \tag{B.22}$$

Applying (B.21) to the above we have

$$h_i(t) = 2\,\mathrm{Re}\,A_i e^{p_i t} \tag{B.23}$$

Writing A_i in terms of magnitude and phase

$$A_i = |A_i| e^{j\theta_i} \tag{B.24}$$

and p_i in terms of its real and imaginary parts

$$p_i = \sigma_i + j\omega_i \tag{B.25}$$

then (B.23) can be rewritten

$$h_i(t) = 2\,\mathrm{Re}|A_i| e^{j\theta_i} e^{(\sigma_i + j\omega_i)t} \tag{B.26}$$

To put the above into real form we need to use Euler's formula, which states that any complex number can be represented equivalently in polar or rectangular form

$$Ve^{j\phi} = V\cos\phi + jV\sin\phi \tag{B.27}$$

Using this on (B.26), we get the impulse response in real form

$$h_i(t) = 2|A_i| e^{\sigma_i t} \cos(\omega_i t + \theta_i) \tag{B.28}$$

Example B.3: As an example, take

$$H(s) = \frac{3s^2 + 22s + 29}{(s+1)(s+2+j3)(s+2-j3)} \tag{B.29}$$

Using (B.4) to find the partial fraction expansion, we obtain

$$H(s) = \frac{1}{s+1} + \frac{1+j2}{s+2+j3} + \frac{1-j2}{s+2-j3} \tag{B.30}$$

In finding the numerator coefficients of this partial fraction expansion, only two had to be found, the last being the conjugate of the previous. Using (B.5) the impulse response $h(t)$ is found

$$h(t) = e^{-t} + (1 + j2)e^{-(2+j3)t} + (1 - j2)e^{-(2-j3)t} \tag{B.31}$$

According to (B.23), the last can be written in the form

$$h(t) = e^{-t} + 2\,\mathrm{Re}[2.24e^{j1.11}e^{-(2+j3)t}] \tag{B.32}$$

Now Euler's formula (B.27) is used to obtain the final real form

$$h(t) = e^{-t} + 4.47e^{-2t}\cos[3t + 63.4°] \tag{B.33}$$

■

In this example we had a pole located on the negative real axis of the s plane, at $s = -1$, and a pair of complex conjugate poles located in the left half of the s plane at $s = -2 \pm j3$. The real pole gave rise to a decaying exponential, whereas the complex conjugate poles gave rise to a sinusoid multiplied by a decaying exponential. Had the real part of the complex conjugate pole pair beeen zero, which is to say that these poles would be located on the $j\omega$ axis, then the impulse response $h(t)$ would have contained a pure sinusoid, and this response would have been representative of an unstable network, one whose power output continues indefinitely. A worse situation occurs when there are complex conjugate poles on the $j\omega$ axis of multiplicity greater than unity. Then the impulse response would contain a sinusoid multiplied by a polynomial, and this rising output is very distinctly representative of an unstable network. To have a stable output the poles must be located in the left-half plane, so that the exponential decay will dominate over the rising polynomial.

We are now in a position to state the rule for stable networks, which is: "*The transfer function poles of stable systems must lie in the left half of the s plane.*"

Some texts on the subject of stability consider poles on the $j\omega$ axis to be conditionally stable. The stricter point of view will be taken here that poles on the $j\omega$ axis are undesirable, and a system possessing such poles is unstable.

Complex Variable Review [3]

Analyticity. $F(s)$ is a function of the complex variable s, and can be written in terms of a real part $X(s)$ and an imaginary part $Y(s)$

$$F(s) = X(s) + jY(s) \tag{B.34}$$

$F(s)$ is said to be analytic in any part of the

$$s = \sigma + j\omega \tag{B.35}$$

plane, if it satisfies the Cauchy–Riemann conditions

$$\begin{aligned}\frac{\partial X}{\partial \sigma} &= \frac{\partial Y}{\partial \omega}\\ \frac{\partial X}{\partial \omega} &= -\frac{\partial Y}{\partial \sigma}\end{aligned} \tag{B.36}$$

The above conditions guarantee that the derivative is unique and is independent of the direction of the differential change in s.

Singularities. Those points at which $F(s)$ fails to be analytic are considered singular. A singularity of great importance is the pole, which is a point at which the function goes to infinity. To ascertain the degree (multiplicity) of the pole we use the pole removal procedure as follows: $F(s)$ has a pole of order (multiplicity) m at s_0 if

$$\lim_{s \to s_0} \left[(s - s_0)^m F(s)\right] = \text{finite} \neq 0 \tag{B.37}$$

Example B.4: For the function

$$F(s) = \frac{1}{1 - e^{sT}}$$

it is suspected that poles exist at $s = jn2\pi/T$ for integer values of n, since the function goes to infinity at those points. It would be interesting to ascertain the order of the poles. The pole removal procedure (B.37) is used (along with L'Hospital's rule)

$$\lim_{s \to jn2\pi/T} \left[(s - jn2\pi/T)F(s)\right] = \lim_{s \to jn2\pi/T} \left[\frac{s - jn2\pi/T}{1 - e^{sT}}\right] = -1/T \tag{B.38}$$

Since the above result is finite we conclude that $F(s)$ has first order poles at $s = jn2\pi/T$. ■

Residues. If $F(s)$ has a pole of order m at $s = s_0$, then it is possible to expand $F(s)$ in the vicinity of s_0 into the Laurent series

$$\begin{aligned}F(s) = {} & \cdots + A_2(s - s_0)^2 + A_1(s - s_0) + A_0\\ & + \frac{B_1}{s - s_0} + \frac{B_2}{(s - s_0)^2} + \cdots + \frac{B_m}{(s - s_0)^m}\end{aligned} \tag{B.39}$$

The constant B_1 is the residue of $F(s)$ at s_0.

The evaluate the residue of $F(s)$ at the mth-order pole s_0, use

$$\operatorname{Res}[F(s)]\big|_{s=s_0} = \frac{1}{(m-1)!}\frac{d^{m-1}}{ds^{m-1}}\left[(s - s_0)^m F(s)\right]\big|_{s=s_0} \tag{B.40}$$

If $F(s)$ is a rational function of s, namely, it can be written as a quotient of two functions

$$F(s) = \frac{N(s)}{D(s)} \tag{B.41}$$

and the pole at s_0 is simple, then the simpler formula can be used for finding the residue

$$\text{Res}[F(s)]|_{s=s_0} = \left.\frac{N(s)}{D'(s)}\right|_{s=s_0} \tag{B.42}$$

Residue Theorem. Let C be a closed contour within and on which $F(s)$ is analytic except for a finite number of finite multiplicity poles inside of the contour C. The integral around the contour C taken in a *counterclockwise* direction can be evaluated in terms of the residues within the contour C,

$$\oint_C F(s)\, ds = 2\pi j(\text{sum of the residues within } C) \tag{B.43}$$

Example B.5: Take the function

$$F(s) = \frac{5s - 4}{s(s-2)}$$

There are two first-order poles, one at $s = 0$ and one at $s = 2$. The residues are found by using (B.40) or (B.42). The result is

$$\text{Res}[F(s)]|_{s=0} = 2$$
$$\text{Res}[F(s)]|_{s=2} = 3$$

For the contours shown in Fig. B.1 the integrals are

$$\oint_{C_1} F(s)\, ds = 2\pi j\, \text{Res}[F(s)]|_{s=0} = 2\pi j(2)$$

$$\oint_{C_2} F(s)\, ds = 2\pi j\{\text{Res}[F(s)]|_{s=0} + \text{Res}[F(s)]|_{s=2}\} = 2\pi j(2 + 3) \quad \blacksquare$$

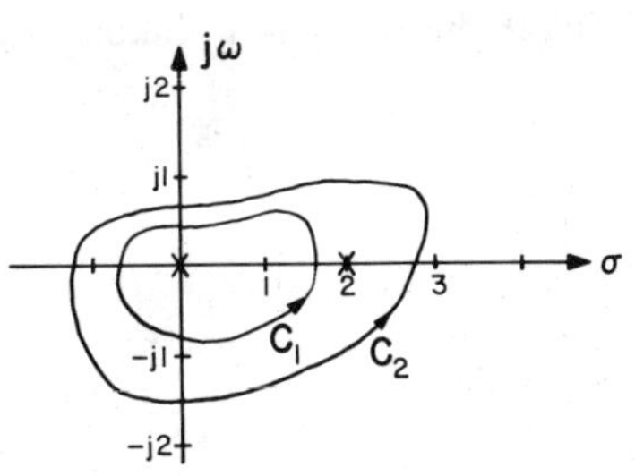

Figure B.1 Two contours of integration in the complex s plane.

REFERENCES

1. Van Valkenburg, M. E.: "Network Analysis," Prentice-Hall, Englewood Cliffs, New Jersey, 1974.
2. Cheng, D. K.: "Analysis of Linear Systems," Addison-Wesley, Reading, Massachusetts, 1959.
3. Churchill, R. V., and J. W., Brown: "Complex Variables and Applications," McGraw-Hill, New York, 1976.

EXERCISES

B.1. In each case, find the partial fraction expansion representation, solve for the impulse response $h(t)$, and if appropriate write it in real form.

(a) $$H(s) = \frac{(s+4)(s+7)}{(s+1)(s+2)(s+3)(s+5)}$$

(b) $$H(s) = \frac{1}{(s+1)^3(s+3)}$$

(c) $$H(s) = \frac{1}{(s+1)(s^2+2s+2)}$$

(d) $$H(s) = \frac{s^2}{(s^2+1)^2}$$

(e) $$H(s) = \frac{1}{s^2-2s+5}$$

B.2. A counterclockwise contour integral encircles the poles at $s = -1$ and $s = -2$ for the integrand

$$F(s) = \frac{(s+4)(s+7)}{(s+1)(s+2)(s+3)(s+5)}$$

(a) Find the value of the integral.
(b) Find the value of the integral if only the pole at $s = -1$ is encircled.
(c) What is the result in parts (a) and (b) if the integral is taken in a *clockwise* direction?

B.3. A counterclockwise contour integral encircles the pole at $s = -1$ for the integrand

$$F(s) = \frac{1}{(s+1)^3(s+3)}$$

Find the value of the integral.

B.4. A counterclockwise contour integral encircles both complex poles for the integrand

$$F(s) = \frac{1}{(s+1)(s^2+2s+2)}$$

Find the value of the integral.

B.5. Find the residues at all the poles for the following rational functions of s.

(a)
$$F(s) = \frac{s^2}{(s^2+1)^2}$$

(b)
$$F(s) = \frac{1}{s^2-2s+5}$$

Index